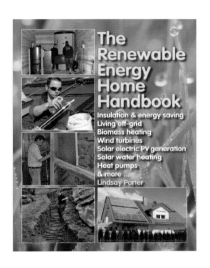

The Renewable Energy Home Handbook

Insulation & energy saving
Living off-grid
Biomass heating
Wind turbines
Solar electric PV generation
Solar water heating
Heat pumps
& more ...
Lindsay Porter

VELOCE

www.veloce.co.uk

First published in January 2015 by Veloce Publishing Limited, Veloce House, Parkway Farm Business Park, Middle Farm Way, Poundbury, Dorchester DT1 3AR, England. Fax 01305 268864 / e-mail
info@veloce.co.uk / web www.veloce.co.uk or www.velocebooks.com. ISBN: 978-1-845847-59-3 UPC: 6-36847-04759-7 © 2015 Lindsay Porter and Veloce Publishing. All rights reserved. With the ex-
ception of quoting brief passages for the purpose of review, no part of this publication may be recorded, reproduced or transmitted by any means, including photocopying, without the written permission
of Veloce Publishing Ltd. Throughout this book logos, model names and designations, etc, have been used for the purposes of identification, illustration and decoration. Such names are the property of
the trademark holder as this is not an official publication. Readers with ideas for automotive books, or books on other transport or related hobby subjects, are invited to write to the editorial director of
Veloce Publishing at the above address. British Library Cataloguing in Publication Data – A catalogue record for this book is available from the British Library. Typesetting, design and page make-up all
by Veloce Publishing Ltd on Apple Mac. Printed in India by Replika Press .

For post publication news, up-
dates and amendments relating
to this book please visit www.
veloce.co.uk/books/V4759

The Renewable Energy Home Handbook

Insulation & energy saving
Living off-grid
Biomass heating
Wind turbines
Solar electric PV generation
Solar water heating
Heat pumps
& more ...

Lindsay Porter

Contents

PUBLISHER'S NOTE

The types of technologies described in this book and the energy savings they provide are relevant in all temperate zone countries of the world. However, it is difficult, and would be unreliable, to specify product manufacturers, professional bodies, quality control organisations and legal standard requirements outside of the author's UK/Northern Europe area of practical experience. Therefore, readers outside the UK will have to do some online searching to find equivalent organisations and relevant legislation in their own country of domicile.

ENERGY COSTS AND PRICING

Wherever the cost of energy, or potential energy savings are referred to in this book, they are shown in kilowatt-hours (kWh). This is an internationally recognised measurement of energy. One kilowatt-hour represents the amount of energy used by (for example) a one kilowatt electric heater running for one hour.

You can easily find out the cost of a kWh of energy in your area by looking on price comparison websites or on utility companies' websites. For instance, texaselectricrates.com, comparethemarket.com.au (Australia) and ukpower.co.uk are three among scores of websites from around the world that show the cost of electricity in their areas – invariably priced in kWh.

In some parts of the world (though, ironically, not in the UK) gas is usually priced in British thermal units (BTU) or in 'Therms' (= 100,000 British thermal units). To convert BTUs to kWh:

1 btu = 0.000293 kWh
1 therm = 29.3 kWh

Alternatively, you could use an online calculator – you'll find lots of them via your favourite search engine.

Introduction
& acknowledgements

Since the early 1980s, I've been writing and publishing technical books – and I hope they've been helpful and friendly technical books – on motor vehicle and associated subjects. Over the same period (and longer) my wife, Shan, and I have designed, built and continuously modified our own house. And always with more than half an eye on conservation and sustainability.

In fact, our first designs for the house where we still live were produced in the 1970s, during which time I spent many hours in Dudley Library, researching building regulations and finding, by chance, the then-current Swedish standards of house insulation. I was amazed, at the time, to discover that the additional cost of building a house to those Scandinavian standards of insulation – roughly equivalent to the UK's requirements for houses built from about 2012 – was scarcely greater than that of building the almost universal 'gas-guzzling' houses. So, we built our house the Swedish way, with not just loft insulation but also floor and foam wall insulation (with a wider cavity between the courses than was normally used); insulating wall blocks and wall boards instead of brick and plasterboard; double-glazing, of

course; insulated pipes and lots more. It was so radical then, but so obvious now.

So, while writing about cars (and I included the relatively eco-friendly fuels we used ourselves such as bio-diesel, LPG power and straight vegetable oil conversions), we were actively thinking and living an earlier version of energy saving.

In recent years, we've gone much further, with a large, solar electricity-generating, photo-voltaic (PV) array, solar water heating, a ground-source heat pump and vegetable oil-powered emergency generator, along with all-LED lighting a range of energy-conserving heat controls ... so, what could be more sensible than to write a book about it all?

Until I started on this book, I thought I knew quite a lot about renewable energy. But when all the research began, I discovered just how much I didn't know. I learned long ago to follow the advice of Robert Persig (author of *Zen and the Art of Motorcycle Maintenance*) – and to encourage others, particularly children, to do so – which is: don't feel bad when you find you don't know something. Be pleased, instead! It's not a sign of failure or inadequacy;

it's a great opening; something to celebrate; a golden opportunity. And so, I'd like to thank the following people for helping me to make the most of my golden opportunity to learn stuff, some of which I didn't even know that I didn't know – if you see what I mean ...

We live near the border between Worcestershire and Herefordshire in central England, and it's been a delight to find several of the top 'renewable' gurus within these two counties. All of them are passionate about their areas of saving CO, saving money, and saving the planet. In order of nearest to furthest from where we live, Simon Holden and Euroheat (HBS) Ltd brings an impressive range of business and educational skills to the field of biomass heating. Paul Hutchens and the crew at Eco2Solar have developed similarly high standards, with special reference to PV electricity generation, as well as solar water heating: installing systems from small-scale domestic right through to large-scale commercial premises. Simon Watts at Eco-Nomical concentrates on domestic installations, including those with a more hands-on approach. The Matthews family at the Efficient Energy Centre is an expanding company, concentrating on bio-fuel and air source heat pump installations, and last, but not least, Leading Edge Turbines combine high standards of engineering with a refreshing honesty about the limitations as well as the advantages of domestic wind turbines. There are three special features that link all of these guys, as well as all the other specialists I mention here. They're passionate about renewables. They're incredibly well informed. And they're as honest as they come!

Moving further afield from chez nous, Newark Copper Cylinders was a mine of information on what is a specialised subject. Mitsubishi Electric Europe BV was wonderfully knowledgeable and helpful about the whole subject of air source heat pumps, while Mike Freeman of Ice Energy is a really helpful bloke who provided much invaluable advice about the installation of ground-source heat pump systems.

The German companies ATAG and Renusol GmbH, along with Ubbink UK Ltd (founded in the Netherlands), displayed technical virtuosity in their respective fields of solar PV and solar thermal systems and fixings, while, from California, Charles Landau – with his website www.solarpaneltilt.com – has provided an almost consummate combination of theoretical calculations and practical measurements with regard to how best to harness solar insolation for PV arrays.

The extraordinary range and depth of practical knowledge exhibited by the Energy Saving Trust can't be overstated while, at the hands-on level, Screwfix Direct produced an amazing range of fittings and materials on a same-day or next-day basis; Makita UK reinforced my already firmly held belief that its hand power tools are the best in the world, while the same status for consumables undoubtedly goes to Würth, whose products I have sworn by for many years.

In the fields of energy-saving, StormDry, Simply LED and Eco-eye wireless electricity monitors have all proven themselves with top quality products.

I'd also like to thank Rod Grainger and everyone at Veloce Publishing. Rod commissioned my first book right at the start of the '80s (and Rod also built his own house with business partner Jude Brooks – so they know what it's like!), and this book has been honed, developed and produced with great professionalism by the Veloce team. And more than anything, I'd like to thank my wife, Shan. She's been my 'through thick and thin' partner in everything we've done over the decades, and this book is a tribute to her in every way imaginable.

Finally, following the approach of PG Wodehouse, who thanked his step-daughter, Leonora, in similar vein at the start of one of his books, I'd like to thank our five 'rescue' dogs and eight cats, without whose unceasing support and attention this book would have been finished in half the time.

Lindsay Porter

Chapter 1
Introduction to home renewable energy

So, like me, you're interested in renewable energy. What does that make us? Tree-huggers? Fruit cakes? Hippies? Or could it be that we're just smarter than those who still don't 'get' it? And if that sounds just a touch too self-congratulatory – oh, who cares! – and here's why ...

- You can actually make money by installing renewables. You're future-proofing your fuel costs in a world where fuel prices are almost certain to rise. And in many countries, you'll actually be paid by the government to generate power. (Well, we already subsidise other fossil fuel and nuclear power systems via our taxes, so why not? It's a great deal for the taxpayer!)
- Because it's the right thing to do! You're cutting your output of greenhouse gases and helping to make life more bearable – even possible – for generations to come.
- You're investing in energy that has none of the enormous risks associated with nuclear power.
- You're creating energy that can and will be renewed. Oil, coal and gas will run out one day. The wind, the sun, growing plants, human waste, heat in the ground: none of these is likely to cease before Hell freezes over!

Our own house on the day the very first of the renewable energy installations that were to be carried out there began. It's now equipped with a biomass log burner and water heater; solar PV electricity generation; solar thermal hot water; a ground-source heat pump, and a vegetable oil-powered generator. It's still connected to the grid, and electricity is sold back to the electricity company.

The great thing is: renewable energy exists freely in nature and will never run out, unlike fossil fuels. But not only that, there's also the problem that burning fossil fuels creates greenhouse gases. Fossil fuels were formed in prehistoric times, incorporating unimaginably large amounts of greenhouse gases at the same time. As the fuels are burned, these gases are returned to the atmosphere for the first time in millions of years. When those fossil fuels are gone, they're gone. For all time. And maybe we, the human race, will be, too. Climate Change Deniers say it might not be true. They could, against nearly all the evidence, be right, but it's a chance I'm not prepared to take ...

WHAT'S INSIDE THIS MANUAL

In this manual, you'll read about all the major, currently available, renewable technologies and how they are installed by the experts. For health and safety reasons, we advise that almost all of this work be carried out by qualified professionals. And in order to qualify for some government grants and payback tariffs, you'll have to have the work professionally carried out and certified, in any case. But seeing how the work is carried out provides a unique insight into how these systems are installed and how they work in detail.

You'll get a much better idea which, if any, of these technologies are suitable for your property and for your personal circumstances. For instance, a central heating system run by a heat pump requires no more involvement, once installed, than any gas-powered system; perhaps less. However, a ground-source air pump system can require a lot of garden area and a huge amount of disruption, which you may not be prepared to tolerate. On the other hand, a wood-burning central heating system need not require a great deal more than swapping (say) an oil-fired boiler for a biomass one, with some provision for storage of wood pellets, for instance, and it will involve some stocking, moving and lifting of sacks of pellets. And traditional log wood burners involve even more – a lot more – work, as I well know! You'll see it all laid out in this manual.

DOWNSIDES

I recently saw an American online advert claiming, 'Install renewables and get your money back in three years!' That's simply not possible in the vast majority of cases, and while – in Europe at least – it's against the law to tell blatant lies in advertising, advertisers are naturally keener on the upsides than any downsides attached to 'their' technology. In this manual, I've tried to point out the 'Againsts' as well as the 'Fors' and I hope that showing how the work is carried out helps provide a better all-round picture.

BE A CLEVER CUSTOMER

Dave Samuels at Leading Edge Turbines tells me that his aim is to have 'clever customers': those

who understand enough about the subject to be able to contribute to the decisions that have to be made. It's a great aim for any specialist to have – and you should beware the ones who would rather you were kept in the dark!

Some installers seem to find renewable technology difficult to understand – though that certainly doesn't apply to any of the specialists featured here. Some technologies are inherently more complex than others; solar water heating and electricity generation (PV) are among the simpler systems to design. Wind turbines are not inherently complex, but you do need an adviser who is honest and realistic about the potential for your site. Some have been installed in locations where it is impossible for them to work effectively. Technologies that heat the house, whether heat pumps or biomass burning boilers, have the greatest scope for mistakes to be made by the ill-informed: possibly something to do with the fact that these technologies, while by no means new, have only been used large-scale in very recent times, so installers and designers have less accumulated knowledge, while statistical, factual information on what works and what doesn't is still being compiled. So it's essential that you choose your installer well. Don't just go on company size or whether or not they are accredited. Try to find and speak to others who have used the company previously. Do your own homework; find out as much as you can about the technology you have chosen, and, if you find you know more than your potential installer, look elsewhere!

- The better informed you are the more chance you will spot a less able installer.
- When your building is surveyed pre-quote, find out if the surveyor is an engineer (ask for qualifications) or a sales person. If it's the latter, politely ask them to go away again. As you will see from this manual, there's a lot of technical 'stuff' that has to be understood, calculated and assessed. Unless you're a qualified engineer, you can't do it. All most sales people can do is tick boxes.
- Try to buy from an installer who is local or who has a network local to you. If your renewable central

heating system goes wrong, or your electricity generation system fails and you're told, "We won't have an engineer in your area for another three weeks," understandably, you won't be too happy.

In the UK, the Energy Saving Trust, a body set up by the government, is enormously influential, and produces invaluable information on all aspects of energy saving and renewable energy choice and installation. Its 'recommended' logo is applied to a wide range of goods and technologies approved by the organisation. It's a mark you really can trust, and the website is full of excellent general advice on energy saving of all kinds.

UPSIDES

None of the foregoing is intended to sound too negative – it's there so that you, the potential customer, can plan for the enjoyment and satisfaction, as well as the cost-savings, that a well-planned and well-installed renewable system can bring. In fact, make this manual one of your upsides! It can't cover every single facet of every conceivable technology, not within the confines of a single, useful book, but it's intended to give you more than enough to be able to make informed choices.

PAYBACK

In most cases, financial incentives are available for the installation of domestic renewable energy systems. In some parts of the world, all of the major systems covered in this manual can qualify for government assistance, usually in the form of payments made over a period of years and connected with the amount of electricity generated or the amount of heat produced. This can actually make the money spent on a renewable energy system a worthwhile investment in itself.

However, the levels of these payments and the steps you have to take in order to qualify for them can change on a regular basis, in recent years, making it unrealistic to try and provide that information here. Those most keen to provide you with the correct information will be the companies who want to sell you their equipment, and there are websites set up by governments and their agencies around the world where you can find out current and detailed information for yourself.

In order to assess the amount of energy being sold back to the grid, as well as the energy consumed, the electricity supplier has to change the traditional meter for a two-way version. You may have to pay for this.

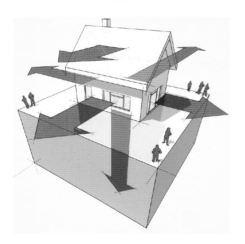

The quickest, most cost-effective and often least disruptive way to save money and cut down carbon dioxide production is to reduce heat loss from your house. We touch on a few ideas in this book, but more details and information on saving energy can be found at the American Council for an Energy-Efficient Economy (aceee.org), and in the UK, the more hands-on Energy Saving Trust (est.org.uk).

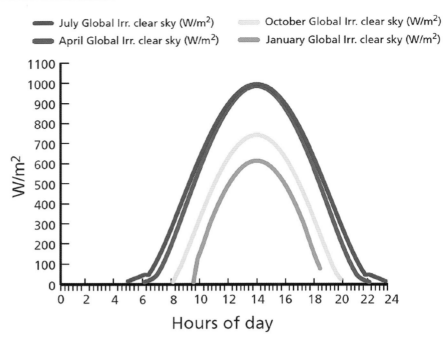

There is an enormous amount of researched information out there to help select the best type of technology. This graph from the UK government's Department of Trade and Industry (DTI) shows the average amount of energy available from sunshine (insolation) in (for example) Manchester, England ...

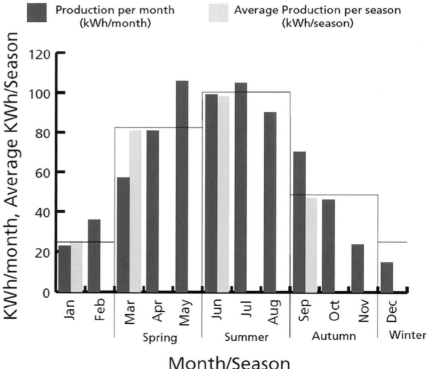

... while this graph shows the average amount by month and by season for the same area.

Chapter 2
Energy saving, insulation and monitoring

ENERGY SAVING

Spending loads of money on renewable technology is not wise if you're not going to make the most of the energy you generate – and it's amazing just how much can be wasted. In fact, saving energy and reducing CO_2 is great for the planet, but it's also great for your bank account. You can save a lot, not just by reducing the amount of energy you use, but also by using it in a smarter way.

Following are some figures that might make you raise your eyebrows in surprise! They have been verified by The Energy Saving Trust (June 2013) (please note: all percentages are per year and costs are subject to change).

You can save 6.4kWh a year (on average) just by turning down the central heating thermostat by 1°C.
(Illustration: Energy Saving Trust)

Where the heat goes (% of your heat loss)

Doors – up to 3%
Floor – 8%
Ventilation and draughts – 12%
Single glazed windows – up to 18%
Roof – up to 26%
Uninsulated walls – up to 33%
Conclusion: insulation saves a packet!

Easy energy saving

Not all energy-saving measures are expensive to put in place: some cost very little, and most are relatively quick and easy to do –

Measure	Potential savings in kWh
Turning down central heating thermostat by 1°C (1°F)	6.4 (3.5)
Avoid leaving electrical devices on standby	5 to 9
Full load in washing machine before use; use the most water and energy efficient settings (wash at 30°C)	1.2
Four-person family replaces one bath a week with a 5 minute shower	1.4. (plus a saving on metered water/ sewerage bills)
Insulate water tank	4.2

Measure	Potential savings in kWh
Switch off lights in rooms not in use	0.7 (average household)
Turn off TV instead of leaving on standby	0.9
Draught-proof doors, windows and letterboxes	2.8
Don't overfill kettle	0.7
Fit radiator reflector panels	0.5
Install chimney draught excluder (open fireplace)	1.8
Insulate water tank and exposed pipes	4.2
Replace halogen spotlights with LED equivalent	2.8
Dry clothes outside when you can	1.8

(Information supplied by the British Co-op, which states: 'In the spirit of always being honest with our customers, we'd like to point out that all savings mentioned in this document are approximate figures, and should only be used as broad guide').
- An energy-saving monitor will increase awareness of electricity use – and save you money!
- Appliance monitors could save up to 15%.
- A smart meter or energy-saving monitor with in-home display can save at least 5%.
- You could recoup the cost within the first few months of use.

Insulation savings (assuming none in place beforehand)

Measure	Potential savings in kWh
Install loft insulation to 270mm (10.62in)	17
Install cavity wall insulation	13
Insulating inside or outside of solid walls	44 to 48
Fit double glazing (secondary glazing)	16.5 (10)

Replacing appliances
In the UK, appliances that carry the Energy Saving Trust or Energy Recommendation logos – A, A+ and A++ – are the most energy efficient choices, while those that are G-rated are the most inefficient. Other countries will have similar organisations/rating systems.
- D-rated fridge/freezer – costs around 9 kWh a year to run.
- Replace G-rated gas central heating boiler with new condensing boiler with heating controls – save up to 30 kWh
- Install a living room thermostat – save around 7 kWh.

Smart heating and hot water tips
- Recommended room temperatures are between 18°C and 21°C (64°F and 70°F).
- Turn down hot water to 60°C (140°F).
- Bleed radiators once a year.
- Set central heating to come on no more than 30 minutes before you return home.

- Close doors and draw curtains/blinds at night.
- Don't cover/block radiators with furniture or curtains.

Fridge freezer savings
- Set your fridge at 0-4°C and freezer at -6°C to -18°C.
- Vacuum your fridge's coils (at the back) twice a year.
- Check door seals aren't broken or filled with food debris.
- Keep frost build-up at less than 6mm (¼ inch).

DON'T WASTE YOUR TIME! Not everything in your home is a major drain on energy. Some things are generally pretty energy efficient, and you'd be better off concentrating on things that matter most. With each of these appliances, 1kWh provides you with this much usage time –

TVs

CRT (old tube-type), 32 inch, 165W	6 hrs
LED, 42 inch, 64W	16 hrs
LCD 42 inch 107W	10 hrs
Plasma 42 inch 195W	5 hrs

Other appliances

Fluorescent strip light (40W)	25 hrs
Stereo	89 hrs
Games console	24 hrs
Laptop	31 hrs
DAB radio	145 hrs
Internet router	159 hrs
LED bulb (6W)	165 hrs
Mobile phone/ MP3 charger	286 hrs

For more information on the topics covered here, visit www.cooperativeenergy.coop

Home efficiency

There's a lot you'll need to assess about your home's energy consumption and insulation before embarking on the installation of any renewable energy system. Indeed, it is a requirement in some countries that, before the approval of any grant or other financial assistance, a house or building has to comply with a certain standard of insulation and energy-saving. Regardless of whether or not this applies to you, it makes sense, in terms of reducing your expenditure and your carbon footprint, not to waste energy.

INSULATION

The most commonly used forms of home insulation are loft insulation and cavity wall insulation. The UK's Energy Saving Trust has invaluable advice in this respect –

"Heat rises and, in an uninsulated house, a quarter of your heat is lost through the roof. Insulating your loft, attic or flat roof is a simple and effective way to save that waste and reduce your heating bills – you can even do it yourself. Loft insulation is effective for at least 40 years, and it will pay for itself over and over again in that time.

"Insulating between the joists of your loft will keep your house warmer but make the roof space above colder. Pipes and water tanks will be more likely to freeze, so you will need to insulate them. If your water tanks are some distance from the loft hatch, you will also need something to walk on for safe access. The cooler air in your insulated loft could mean that cold draughts come through the loft hatch. To prevent this, you can fit an insulated loft hatch and put strips of draught-excluding material around the edges of the frame."

Cavity walls were built into most – though by no means all – homes from around 1900. The cavity brick-built wall forms an insulating air gap, and, although it's not a very efficient one, it's a lot better than a solid wall. The Energy Saving Trust once again states, 'Did you know that around a third of all the heat lost in an uninsulated home

goes through the walls? Heat will always flow from a warm area to a cold one. In winter, the colder it is outside, the faster the heat from your home will escape into the surrounding air [through the walls].'

There are several insulation options for buildings with solid walls, including –

- Fitting a new, foil-backed, plasterboard wall surface on all external walls, mounting the plasterboard on battens to provide a narrow air gap.
- Fitting a new, secondary wall made from insulating materials and finished with weatherproof panelling on the whole of the outside of the building.
- Adding insulation to the inside of an outside wall, as shown in the accompanying images.

ENERGY SAVING

As the Centre for Sustainable Energy, which serves people living in the Bristol and Somerset area states, 'Some

So that the sheets lay continuously, without gaps and also in order to increase strength, we used Würth Bond and Seal, a PU sealant that works particularly well with PUR/PIR rigid foam insulation.

The 18 inches thick walls in the original, stone-built part of our house consisted of two courses of random-sized stone filled with rubble. We decided to use rigid foam insulation, and builder Matthew began by cutting the foam to shape, then placing plaster dabs on the wall in order to hold the insulation in place.

With the plaster dab still soft and sticky, Matthew placed the pre-cut sheet of foam in position ...

... and used a straight, relatively heavy piece of timber to tap it down onto the plaster dab, ensuring that it was completely level with the sheet or sheets already fitted.

Both of the outside-facing bathroom walls were covered in this way, with the foil side facing inward to reflect heat back into the room. Later, we used plastic tongue-and-groove board to finish the wall, but you could use almost any finishing material of your choice.

The other major problem with solid walls is that they allow driven rain and dampness to penetrate from the outside. This causes several problems: some obvious; some less so. For example, when wet brick, stone or concrete freezes, the surface of the wall breaks away in a natural process of erosion. What also happens is that, as the moisture in the wall evaporates, the wall itself and the building inside it are naturally cooled – it's a process by which a simple, non-energy consuming coolbox can be made to work, by the way. So, what's to be done? Well, the first step is to repair and re-point

the wall, as necessary. Once again, there are many technical matters to go into, particularly when replacing lime mortars, that there isn't room to go into here, but suffice to say that the wall needs to be sound to start with.

If you need to check the porosity of your wall, you can buy a simple gauge which is stuck to the wall and topped up with water …

… with several of these gauges placed on the surface of the wall, the amount of fluid absorbed can be observed and compared with the acceptable or expected absorption levels described in the data sheet supplied with the gauges. (Illustrations courtesy StormDry)

Silicone is frequently used for attempting to make absorbent walls waterproof but this is a strictly temporary measure which may last no more than 12 months at most. StormDry wall treatment, while very much more expensive, lasts for many years. We found it to be extremely beneficial – so much so that it has also completely stopped water penetration around several exposed windows. It allows moisture from within the wall to evaporate, but does not allow water to penetrate from the outside-in. In fact, it can be seen to work quite clearly by the water that beads on the surface of the wall before running off.

electrical items use a lot of electricity. Others don't. As a rule, those with moving parts or which produce heat use much more than those producing light or sound. So if you want to save electricity and money, there's no point worrying about a digital clock or an electric razor, since these use so little power you would hardly notice the difference. The big savings lie elsewhere.'

And if you want to know how much electricity you're consuming, 'Every electrical appliance has a power rating which tells you how much electricity it needs to work. This is usually given in watts (W) or kilowatts (kW). Of course, the amount of electricity it uses depends on how long it's on for, and this is measured in kilowatt-hours (kWh).'

In a typical house, 10 to 20% of energy consumption goes into room lighting. CFL bulbs, which are in effect mini, twisted versions of striplights, are great for saving money on lighting, and you can even buy efficient outdoor lights that cost a fraction of the price of halogen floodlights to run.

We have fitted LED (light-emitting diode) lights throughout our house. LEDs are simple, solid-state electronic devices that produce a small amount of light. Domestic LED bulbs contain lots of LEDs so that a bright enough light is emitted.

LED like-for-like replacements are still, at the time of writing, expensive, but they are the most efficient type of bulb and pay for themselves several times over before they need replacing – provided you don't buy cheap and nasty bulbs to start off with. We tried some; they turned out to be very expensive and potentially dangerous: none of them lasted long enough to be cost-effective, and two of them actually exploded …

In the UK, look for the 'Energy Saving Trust Recommended' label on LED bulbs and fittings to ensure you buy the most energy-efficient products on the market.

MONITORING

Right at the very start of the planning process it's essential you know how much power you could

First-generation LEDs (inset) give out relatively little light and require a large number to be useful. Second-generation LEDs (left) give out more light per unit, while third-generation bulbs are even brighter: although more expensive, fewer LED units per light fitting are required.

LED bulbs are usually available in one of two colour temperatures: the two shown here on the left are generally regarded as 'soft' light. They are undoubtedly easier on the eye but produce less visible light than the whiter type of bulb shown on the right.

The Power Predictor 2, from Eco Ark Ltd, is a wind and solar power predictor, which calculates how much electricity you could generate if you installed solar panels or a wind turbine at your location. It includes a wind vane with balanced potentiometer for measuring prevailing wind direction, and a three-cup anemometer, said to be accurate to +/- 3% in independently verified tests.

The Power Predictor 2 allows you to collect solar and wind data to assess site potential. There is a self-contained, low-power, waterproof data logger (SD memory card and USB adaptor included). Its LCD screen provides live indication of windspeed and solar energy. The device logs data, which can then be uploaded to your own online account. Data is transferred to your computer by temporarily removing the SD card from the Power Predictor data logger. The solar PV cell records solar irradiation over a period of time, though my only criticism is that it was found to be over-pessimistic, since the Power Predictor was installed after our domestic PV panels had been fitted. This is much better than being over-optimistic in this regard, however!

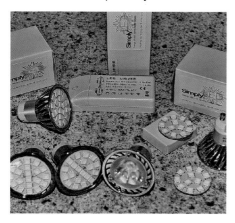

LED lights seem to be evolving all the time. Ours, from Simply LED, consist of second- and third-generation bulbs, although we also have a few of the less-bright, first-generation bulbs in our house; still going strong after several years. Bulbs can be purchased for almost all fitting types, including, as seen here, the flat-type for lights which are located beneath kitchen cupboards. These not only save a good deal of money compared to the original halogen fittings, they also run almost completely cold, removing the risk of burning or scorching commonly experienced with halogen types.

generate before planning to invest in renewables. Different technologies may or may not be suitable in specific locations and, in the worst case scenario, you may even find that none of them will be suitable for you. However, a small amount of research early on could save you from wasting a small fortune.

Among the more obvious technologies are biomass furnaces and air source heat pumps, while the technology requiring most background investigation is almost certainly the one that deals with wind turbines – see the relevant section of this book.

Power Predictor's web application works through your browser, on either PC or Mac. You can run your own power report using the software accessed through your account, and so discover which of the wind turbines and solar panels on the market are best suited to your home, and how long they would take to pay for themselves. The price includes a two-year site license valid for one main site and two sub-sites.

In the majority of cases, Eco-Eye monitors are simple to install – no need for an electrician. However, if you need to open up any electrical equipment – as in the case of our ground-source heat pump's electrical input, which was entirely contained inside a control box – you may need to call on the services of an electrical engineer.

… which then instantly provided a reading to the Eco-Eye wireless receiver/display, which we chose to mount where it could be easily seen in a living area. You can see that on this occasion, the heat pump was consuming approximately 3.25kW, which is just over one third of its 9.2kW maximum output. Very occasionally, consumption rises nearer to 4kW but, more often, especially after it's been running for a couple of hours, it drops to between 2.8 and 3kW, indicating that our ground-source heat pump is operating well and consistently throughout the winter season, with an acceptable CoP of around 3.0. (See Chapter 7 for details of what this means.)

Almost as important is the need to keep your eye on power consumption both within the home and by the heat pump, if you have one fitted. There are several different types of electricity monitor, and I tested those from several manufacturers. The ones that produce the best results in our applications are Eco-Eye wireless electricity monitors. Eco-Eye produces a range, from the simple Eco-Eye Plug-in individual appliance monitor to the sophisticated Eco-Eye Smart with computer connectivity options. Whichever monitoring system you use, you will be better informed about where and how you can make savings in your electricity consumption. There's no point, for instance, buying a timer for turning on and off an extremely low-power electrical component, because the cost of the timer could greatly exceed any potential saving. On the other hand, you might find that a component such as a large computer or satellite TV box consumes more electricity than you realised, in which case, you will be sure to turn it off whenever you can! The only way to be sure, is by monitoring.

Jeremy, our electrician, connected the monitor's pickup and, using the LED supplied, turned on the power and checked that it was working.

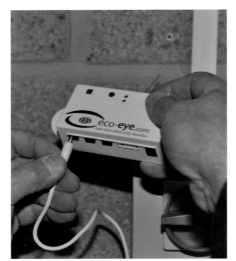

The cable was then connected to the Eco-Eye wireless sender unit …

Placed next to the Eco-Eye in our kitchen is an SMA Sunny Beam, which wirelessly transmits the amount of power that our SMA PV system is generating. You can set it up to show many different parameters, and can also transfer the data to your computer to keep a record of long-term power production. This was recommended to us by one of our installers, Eco2Solar, and has turned out to be an absolutely invaluable way of keeping an eye on power production, and on any faults that may occur such as through part of your power circuit temporarily going down, which will, if it's on the same circuit, simultaneously disable your PV inverter. It can collect (aggregate) the total generated through several inverters or show individual inverter output.

In order to qualify for UK government financial incentives, it is necessary (at the time of writing) to have a 'Green Deal' assessment carried out on your home to ensure that it is sufficiently energy-efficient to start with. Sue Cole, a qualified assessor working on behalf of Eco2Solar, carried out our checks. Having a Green Deal Advice Report produced is a very useful exercise for anyone new to the idea of energy saving, and it provides some useful advice (as well as some computer-generated inanities, but that's life ...). It's a good starting point for energy savings.

Chapter 3
Off-grid and grid-tie with battery back-up

The vast majority of those carrying out renewable energy projects wish to remain connected to the electricity grid. However, a smaller number – either because they have no alternative, their electricity supply is unreliable, or simply out of choice – may need to know how to live 'off-grid' with a true, stand-alone, off-grid system.

BATTERY-LINKED, ON-GRID SYSTEMS

However, a more recent trend – promoted by a small number of MCS-accredited companies such as Bright Green Energy Ltd – is for the home to be both grid-linked AND fitted with storage batteries. This has two main benefits –

- Surplus energy from solar PV or wind turbine renewables is stored in the batteries when not being consumed. The power is then drawn down from the batteries when required, saving the cost and CO_2 of extracting that electricity from the grid.
- If there is a power cut, the system can be connected so that lower-consumption equipment in the house, such as lights, computer and TV (but not including washing machines and heating, etc) draws

power from the batteries. What is more, the batteries can continue to be charged from the renewable resources through a second inverter. (This can't be done with a regular, grid-tied inverter because PV and wind turbine inverters have to isolate themselves from the grid when no grid power is detected.)

A setup incorporating batteries is best considered (if at all) when a new renewables generation system is being installed. This is because it requires a different type of inverter to the one normally used, and, since the cost of any grid-tied inverter is quite substantial, it makes the project more cost-effective to install an inverter capable of being battery-linked right at the start.

There are, of course, downsides to having a battery-linked system, and one is that you will need to safely store the batteries (which will require a certain amount of maintenance). The main drawback, however, is that of cost. If having a battery-linked installation in order to cut the amount of electricity you need to purchase from the grid, consider carrying out a cost-benefit analysis first. In other words, will the saving in electricity be greater than

the cost of the initial setup, and that of replacing lead-acid batteries (say) every five years?

There's a good deal more information on battery-linked, on-grid systems later in this chapter.

OFF-GRID SYSTEMS

The following is an extended version of information available from the Energy Saving Trust, and information published by the UK's Department of Trade and Industry (DTI).

Isolated homes with no mains electricity supply either have to make do without electricity, or generate their own. For these residences, a renewable electricity generation system – using wind, water or solar power to generate power – could provide a solution. Living with an off-grid system is never the same as being on mains supply, but it can be cheaper than getting a mains connection, and is much cheaper and quieter than running a diesel generator. A hydro resource may be reliable enough for you to do without batteries or back-up generator, but this depends very much on the site (and on the needs of the household): suitable sites – and those allowed by the authorities – are few and far between.

Minimising power use

The first step toward setting up an off-grid renewable system is to minimise your home's electricity usage. This is important for all home renewables systems, but if you're off the grid it's vital. To reduce the amount of electricity you need – and therefore the size and capital cost of any renewable energy systems for electricity – tasks such as water and space heating can be accomplished using dedicated renewables such as biomass boilers or solar water heating.

Surplus wind- and PV-generated electricity can be used to complement water heating systems by putting 'unused' power into an immersion heater, once appliance demand is met and your batteries are full – though there's rarely enough spare power to cater for all water heating needs.

Converting battery voltage to 230V (or whatever your mains voltage is) is achieved by running the low-voltage electricity through an inverter. But an inverter itself consumes power – probably somewhere around 10%, and up to 20% in the worst cases – which is valuable power going to waste if it's not absolutely necessary to run an inverter.

To save that energy loss, you may also consider using lighting and other appliances, such as fridge, TV, satellite, iron, hairdryer – there's a huge list – which run from battery voltage (either 12V or 24V). Some of these appliances are intended for use in caravans or on boats, so may not have the capacity to do a 'proper' job on a permanent basis. You could search through off-grid living or boating forums, for instance, for the experiences of those who live with low voltage on a daily basis. Of course, if you wanted some appliances to run off 230V and others off 12V or 24V, it would mean having to run two separate (and safely separated) wiring systems, in accordance with local electrical regulations.

Water heating

If you have a biomass heater (such as a wood stove), a back boiler is a fine way of heating domestic water, especially if you also have a solar thermal water heating system on the roof (this will need a pump, although there's nothing to stop you adding an extra solar PV panel (if necessary) and running a low-voltage pump – see Chapter 4. The advantage is that the pump is usually only needed when the sun is heating the water, at which time it's almost always providing pump power, too.

Room heating and cooking

Electric cookers, space and water heaters all use a very large amount of electricity. For instance, electric cookers use about 1.5kW per hot plate, so LPG and solid fuel stoves are usually preferred for heating and/or cooking. Microwave ovens draw a lot of power, but since food is cooked much more quickly, total power consumption may not be excessive. There are some low voltage microwaves available, though these cook more slowly.

Lighting

The most efficient in terms of low power consumption and high light output is LED lighting, though it is more expensive to buy. Otherwise, compact fluorescent lights are almost as efficient – check the figures for yourself! Incandescent bulbs are dreadfully inefficient, even in their low-voltage versions (they're not usually available for mains voltages) although regular fluorescent tubes are a reasonably efficient halfway house.

Refrigeration

LPG-powered fridges are a good choice, provided suitable gas cylinders are available. Many fridges, having been produced with the leisure industry in mind, are capable of running on LPG, 12V or 230V – whichever is available or more convenient. These are known as 'three-way' fridges.

Battery-linked, on-grid systems

(The following advice on generators was supplied by Bright Green Energy)
In most cases, off-grid power systems have seasonal load requirements, and are designed around components that usually result in the deployment of an over-specified system, because there may be discrepancies between seasonal loads and sunshine and/or wind availability. In such cases, an alternative source of energy will be required to handle load mismatching, and as a backup to the renewable energy source, in which case it is common to install either a petrol, diesel or propane AC generator as a source of backup power. Bright Green Energy strongly recommends against using DC generators for mains-powered installations.

Types and sizes of generator

The selection criteria for a generator can be whittled down to –
- Initial cost (but do think long-term!).
- Power output (it must meet load demand).
- Fuel availability and the local cost of different fuels. (See Chapter 9 for details of vegetable oil-fuelled generators.).
- Noise and vibration – air-cooled diesels can be much too loud, especially if you have neighbours.
- Maintenance schedules and ease of access for maintenance. Poor maintenance design can make draining oil, for instance, impossible without random oil spills.

Generator characteristics

These include: speed, efficiency, fuel type, frequency, noise and vibration levels, overload, waveform output (true sine wave), and starting capability.

Rotation speed

60Hz generators operate at either 3600 or 1800rpm; 50hHz generators operate at 3000 or 1500rpm.

Units operating at 3000 or 3600rpm are two-pole machines, while those operating at 1500 or 1800rpm are four-pole machines. In other words, while both generate the same amount of electricity nominally, the four-pole machines do so at half the rpm, which is important because the higher rpm is, the more wear and tear on the machine the noisier it will be. Four-pole machines are more expensive, but are recommended when larger size, heavier-duty units are required, and when more than 400 hours of operation time per year are anticipated. Two-pole machines are used in lighter-duty applications where under 400 hours' operation per year is anticipated.

A generator operates most efficiently when run at full capacity so, ideally, should be run flat-out for short periods. Obviously, the size of the generator should take into account likely peak power consumption of all the appliances that it needs to power. A small generator (<3kW) will be fine if you can be disciplined about appliance

It would very difficult indeed to function in a normal, modern way without having a generator installed. It doesn't need to excessively pollute the environment, however – see Chapter 9 for information on vegetable oil-powered generators. It will usually be necessary to restrict the use of the heaviest appliance loads (washing machine, vacuum cleaner, etc) to when the generator is running, and use any spare generating capacity for battery charging or water heating. This **SDMO Single Phase 110V/230V Diesel Generator is connected in conjunction with a battery bank at an LE Turbines off-grid installation.**

usage ("No, darling, NOT the toaster!") but a larger, say 7.5kW, generator will suport several appliances at once – but watch out for fuel consumption! Ideally, vacuum cleaning and washing should be carried out when weather conditions are windy or sunny, which would mean that generated power was being used directly, which is fine if you've got the time and energy to organise yourself in this way. But it's a big commitment!

Generator selection

When we selected our own 10kva backup generator (capable of producing 8-to-9kW or more), we looked for a number of key points.

- It had to have a water-cooled diesel engine (so that it has the potential for Combined Heat and Power (CHP) usage in future, where the heat from the engine coolant is used in the home instead of being wasted to atmosphere.
- The engine had to be a reputable make, running at 1500rpm to reduce wear and noise.
- The capacity had to be enough to power most regular household items, including our 3kW+ ground source heat pump, if necessary.
- The generator (alternator) also had to be a reputable, well-known make. Unbranded Chinese-made units, can mean having the devil of a job finding parts (if at all), when necessary.

- It had to have a 'super silent'* sound insulation enclosure.
- Reasonable fuel consumption.

*There is literally no such thing as a 'silent' – or even a 'quiet' – diesel generator. What a confusing term! All you can go on is the manufacturer's published figures in decibels (dB) – but make sure you are comparing like-with-like, and that the dB figures are all measured in consistent formats and distances.

Noise in decibels (dB)

The decibel scale is not linear, unlike a ruler, where the difference between graduations represents a constant measurement. The decibel scale is a logarithmic scale where the difference between each mark is a multiple of 10. So, the scale goes from 1 to 10 to 100 to 1000, and so on. Therefore, a doubling in sound level does not double the number of decibels recorded.

The following are some common sounds against which you can compare the quoted dB of a given generator – though note that diesel generators have a lower-pitched sound and low-pitched sounds are (generally) less annoying than high-pitched ones – but are capable of travelling further – not indicated by the dB reading.

Sound level (dB)	Circumstances
140	Threshold of pain (134dB)
120	Loud nightclub, standing at speaker (120dB)
100	Pneumatic drill at 5m (100dB) or heavy goods vehicle, from pavement
90	Powered lawnmower at operator's ear
80	Average traffic on street corner (74dB) or vacuum cleaner at 3m
70	Telephone ringing at 2m or conversational speech
60	Typical business office (54 dB)
50	Living room in suburban area or refrigerator humming at 2m
40	Library (34dB)

This Genset-made unit has one of my favourite engines, built by Kubota – but the lower-revving 'traditional,' 3-cylinder Kubota unit – the very well proven, well-known Italian alternator and control panel manufacturer. Single-phase. 79dB and water-cooled. 2.2 litres/hour at 75% load. Here, it is being connected to an ATS, using wireless remote switching, by Victor Samsonic, whose company, GenControl. co.uk, designs and produces such devices. An automatic transfer switch (ATS) control panel is designed to sense when an electric mains outage occurs, and then activate the backup generator ...

Automatic transfer switch

... when the ATS recognises that the generator is ready to provide electric power:

1. The ATS breaks the home connection to the electric utility (to ensure the ...

... generator is not connected to the electric utility), and –

2. Connects the generator to the home main electrical panel. When utility (mains) power has been reinstated for a set amount of time:

1. The ATS transfers back to utility power, and –

2. Commands the generator to turn off, but only after another specified period of 'cool-down' time with no load on the generator.

This ATS, in our house, supplied by Victor Samsonic of GenControl.co.uk, also has a manual change-over switch, mounted on the front door. This is so that the generator can be turned on without there being a power outage. The generator supply to the mains utility is then shut off, in the same way as with a power outage.

The GenControl ATS system works in this way. Under normal circumstances, when mains power is available, the utility (grid) power supply runs through the Automatic Transfer Switch (ATS) contactors, and connects to your distribution board. When mains power fails, the GenControl ATS will pause for a 25 second period to ensure there hasn't been a power spike. During this time the 'mains available' LED flashes. The ATS will then initiate a generator start signal, warm up the generator for a time preset by the internal programmable timer, and then connect the generator power supply to your home or business premises. When mains power is restored, the reverse happens and the controller automatically switches back from generator power to mains, shutting down the generator after a cool-down period (again, preset by internal configurable timer), and restoring it to standby mode.

BATTERY SYSTEMS

This section is relevant whether the batteries form part of a true stand-alone (off-grid) system or part of a hybrid (eg grid-linked, battery-linked) system.

The heart of most off-grid systems is the battery store. A bank of deep cycle batteries will store electricity when it is generated, and provide power for when it is needed. Most deep cycle batteries are lead acid batteries, like car batteries but with thicker plates to cope with the deep charging cycle. Batteries with liquid acid in them require care, as they give off hydrogen gas when charging, and may lose water over time, which will need replacing. More expensive AGM batteries do not contain liquid, and are much easier to look after. With proper controls and system design, a battery bank may last five years or more, but it won't last for ever, so you should budget to replace the batteries several times during the life of a system.

- Batteries can be charged by more than one generation technology: you might have an exposed site with a wind turbine, some photovoltaic panels to provide input during windless spells, and a diesel generator as backup and for running high-power appliances for short periods.

- The size of your off-grid storage system depends on personal circumstances. Estimate the size you'll need by multiplying your average daily demand by the number of days you want your battery system to power your home for (called autonomy). This will tell you the useful energy you want your system to be able to supply – though the depth that batteries can discharge to, usually around 50%, will increase the size you need.

- Storage technologies are not limited to deep cycle lead acid batteries, although these are the most widely used. Other types of battery are available such as nickel metal hydride (NiMH) and lithium-ion (Li-ion) versions, though these can be very expensive, and neither is without disadvantages. Other technologies, such as flow batteries, may become commercially available.

Some inverters can act as a battery charger, as well as look after the batteries and controll the system, or you may have several separate 'boxes' for each of these different functions. Peak power consumption should be considered when choosing inverter size. The continuous rating of the inverter should be higher than the amount of electricity you are likely to consume – the degree of safety margin

It is far better if larger battery banks are made up of 2V cells connected in series (such as in this Rolls battery installation) rather than lots of 12V batteries in parallel. This prevents circulating currents within the batteries, and thus extends battery life.

With an inverter to convert low-voltage DC power to mains-type voltage AC power, electricity from the batteries can be used to produce mains voltage alternating current (AC), which can run standard appliances, as well as allow the batteries to run low-voltage lighting, and perhaps other direct current (DC) appliances. This is a Ring Automotive RINV1100 1100W Pro Inverter, capable of 12 hours' constant use at 1100W: an effective piece of kit at a competitive price.

will be your decision. Calculate power usage by totalling the consumption in watts of all those devices that are likely to be used at the same time.

Fortunately, most inverters have an overload capability, for very short periods of time, of two or more times their rated output – check before buying the inverter. This capacity allows the inverter to cope with starting surges from electric motors, which can be as much as three times their rated power. It's essential not to start up too many inverter-driven appliances at the same time.

Charge controller

This provides the regulator/dump interface between the electricity generator such as a PV array or wind turbine and the battery, in order to prevent battery overcharging. The unit may also provide other functions such as maximum power point tracking (MPPT), voltage transformation, load control, and metering.

The charge controller must –
- Be rated for the minimum and maximum current and voltage ratings.
- Be labelled in accordance with regulations for your region.
- In Europe, carry a CE mark.

A full recharge is important for good lead acid battery health. A small-size cable between the charge control unit and the battery – with an associated high-voltage drop – may lead to the control system prematurely halting the charge cycle. These cables should therefore be sized for a maximum voltage drop of less than 1% at peak generator output. For controllers with a separate battery-sense function, a fused battery-sense cable can be installed.

Battery over current protection

A battery stores significant energy, and has the capacity to deliver large fault currents. Proper fault protection must be provided. An over-current device must be installed in all active (non-earthed) conductors between the battery and the charge controller. The length of cable between the over-current device and battery terminals must be as short as practicable.

The over-current device (either a fuse or circuit-breaker) must –
- Have a trip value as specified within the charge controller manual.
- Be rated for operation at dc; at 125% of the nominal battery voltage.
- Have an interrupt rating greater than the potential battery short-circuit current.

Battery disconnection

A means of manual isolation must be provided between the charge controller and the battery, either combined with the over-current device or as a separate unit. The isolator must be double pole, dc rated and load break, while the length of the cable between it and the battery must be as short as possible. Isolation ability is essential and the system must be designed so that the generator cannot directly feed the loads when the battery has been disconnected.

Combined fault protection and isolation –
- A circuit-breaker provided for battery fault current protection may be used to provide isolation, if it is rated as an isolation device.
- A fuse assembly provided for fault current protection may be used to provide isolation if it has readily removable fuses (eg fuse unit with disconnect mechanism).

An ATS Automatic Transfer Switch Panel is able to turn on a generator when a mains power cut takes place while isolating the supply from your generator's power. It can then switch back to mains when the power is restored. You MUST comply with local regulations regarding grid connections.

Cables in battery systems

All cables must have a current rating above that of the relevant over-current device (nearest downstream fuse/circuit-breaker). Cable current ratings should be adjusted using standard correction factors for an installation method: temperature, grouping and frequency to local electrical and wiring regulations.

BATTERY SELECTION

Some key considerations are:
- In the majority of cases a true 'deep cycle' battery will be required.
- It must have an adequate storage capacity and cycle life.
- Will the battery be made up of series cells or parallel banks? While series cells will generally give better performance, practical considerations may influence the design. In general, though, banks with more than four parallel units are to be avoided.
- 12V or 24V? The most commonly used is 12 volts, for which a wide range of lights and appliances are available. However, 24V systems are better where long cable runs are required (because of less voltage drop) or if an inverter of more than 1kW is required. Some 24V lighting and appliances are available, though fewer than for 12V systems.

Battery sizing

Careful attention must be paid to battery sizing to ensure maximum battery performance and longevity. Three factors should be considered when sizing battery banks for renewable energy systems: 1) the electrical power required, or load, of a particular application – which will be affected by climate and seasonal weather patterns; 2) the maximum depth of discharge (DOD) the battery will be allowed to go to (20% minimum for lead acid batteries), and 3) the number of days the battery will be used to power the loads. For a small system, three to four days' storage will usually be sufficient, especially when a generator is available for back-up.

Fortunately, Trojan Battery Co, a manufacturer of deep cycle batteries, has developed an automated battery sizing calculator to accurately size a battery bank for renewable energy applications.

This simplifies the task of properly sizing a battery bank for off-grid renewable energy systems. Trojan says that failing to properly design the system to meet specific load requirements leads to:
- Improper sizing of the batteries, which can reduce the life of the battery bank, and have a dramatic impact on system performance.
- Potential loss of system power.
- A replacement battery being required sooner than necessary.

Trojan's battery sizing calculator is an easier way to determine battery capacity than manually calculating load requirements and then converting them to battery capacity. Customers simply fill in the appropriate information on the electronic form, such as battery voltage, desired depth-of-discharge (DOD), days of autonomy, AC and DC loads, device types with power ratings, and hours per day or days per week used, and the application automatically determines the necessary battery capacity, and recommends the relevant Trojan battery model for that particular setup. The calculator also allows customers to run 'what if?' scenarios to find specific battery options to meet their budget or configuration requirements.

The battery sizing calculator can be found at www.batterysizingcalculator.com.

BATTERY INSTALLATION

Sufficient ventilation is needed to remove battery gases, with an air inlet at low level and an outlet at the highest point in the room or enclosure. This is particularly important in the case of vented lead acid units as hydrogen is given off during charging – and a concentration of greater than 4% creates an explosion hazard.

Ventilation also prevents excessive heat build-up. The ideal operating temperature for a lead acid battery is around 25°C, and temperatures significantly above or below this will lead to reduced lifetime and capacity. Indeed, at very low temperatures, discharged batteries may freeze and burst; at high temperatures, thermal runaway (a repeating cycle in which excessive heat causes more heat until the operation ceases or an explosion occurs) can occur in sealed batteries.

Battery banks must be housed in such a way that:
- They are in compliance with local regulations.
- Access can be restricted to authorised personnel.
- Adequate containment is assured.
- Appropriate temperature control can be maintained.
- Battery terminals are to be guarded so that accidental contact with persons or objects is prevented.
- Items which could produce sparks (eg manual disconnects, relays) should not be positioned within a battery box or directly above one.
- Battery gases are corrosive, so cables and other items inside a battery enclosure need to be corrosion resistant. Sensitive electronic devices should not be mounted in, or above, a battery box.

To ensure proper load/charge sharing in a battery bank made up of units connected in parallel, the units need to have the same thermal environment and the same electrical connection resistance.

In larger battery banks, fusing each parallel unit should be considered.

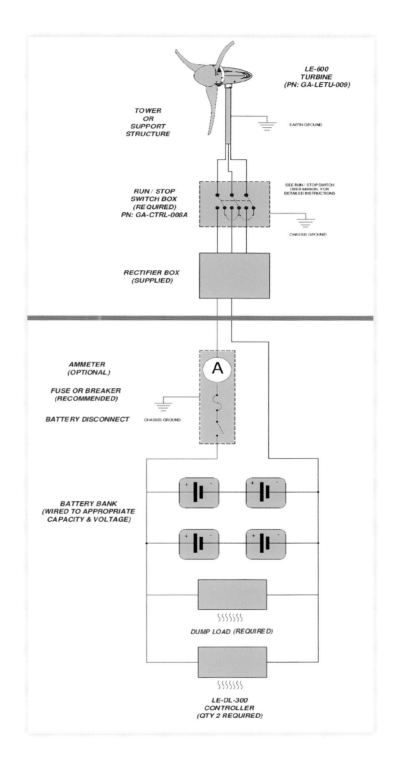

This drawing from LE Turbines shows how a single energy source – in this case a wind turbine (whose components are shown above the red line) is connected to a battery bank ...

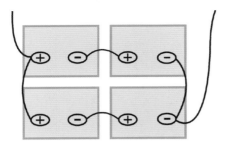

In a typical connection configuration for a small series parallel battery bank, take-offs are on opposite corners ...

LABELLING

Warning signs must always comply with local regulations, but the following are examples of warning signs that may be displayed –

- No smoking or naked flames.
- Batteries contain acid – avoid contact with skin and eyes.
- Electric shock risk – xxx V dc.
- Circuit protection, and all points of isolation should also be labelled with 'dc Supply – xxx V dc' (xxx representing the appropriate voltage).
- Protective equipment, including appropriate gloves and goggles – together with an eye wash and neutralising agent – should be stored adjacent to the battery installation.

All labels should be clear, easily visible, and constructed and fixed so as to remain legible and in place throughout the design life of the system.

ENERGY STORAGE FOR GRID-CONNECTED, RENEWABLE ENERGY SYSTEMS

The following section is based on information from Bright Green Energy Ltd, a company at the forefront of developments in this area.

Bright Green Energy has developed the @HOME battery storage system so that you can install a home battery back-up system and protect yourself from power cuts. Utilising existing technology, the @HOME battery system consists of batteries, inverter charger, battery monitor, and a few minor adjustments to the electrical fuse board. You continue to receive the feed-in tariffs (when applicable) while charging your batteries direct from your renewable energy system.

Adding batteries to your

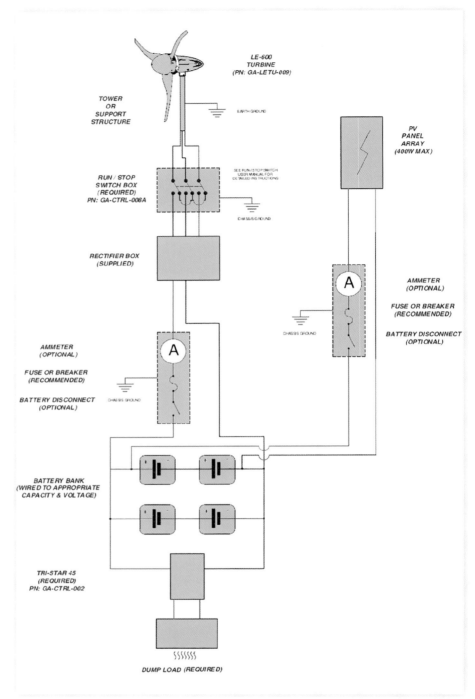

... while this is how one of its multi-source installations – wind and PV in this case – might be connected.

renewable energy system offers several advantages –
1. Reduced export. Charge your batteries for 'free' with unused, self-generated electricity.
2. Reduce your electricity bill still

further by running appliances from batteries during the night when your solar PV system is not working.
3. Secure power supply to vital appliances. In the event of a power cut keep your fridge/freezer operating.

4 Batteries can be added at any time to any existing solar PV or wind turbine system.

The electricity grid and your solar PV system

Much of the following also applies to wind turbine systems: the obvious difference is that, with PV systems, the battery system draws power when the sun isn't shining whereas with wind turbine systems, it kicks in when the wind isn't turning the turbine rotor.

Grid-connected solar PV and wind turbine systems do not work during a power cut. Regrettably, the system remains idle until mains power is restored to the property. Straight grid-connected solar PV systems are 'battery-less.' The solar PV inverter requires a connection to the electricity supply to function properly; if no grid supply is available, the solar PV inverter does not operate. The obvious question is: if a solar PV system produces energy whenever the sun shines, why can't we use that energy even if the grid is not available? The answer is not so clear-cut.

It should be remembered that a solar PV system is a constant current source, and therefore cannot increase or decrease the amount of energy it produces to handle normal, everyday appliances and home load demands. If the solar PV system is intended to produce 8A, it cannot produce more than that. Of course, appliances which operate at 8A or less could be run, but fridges and appliances with motors would not work as they require a higher 'surge' current in order to turn over the motor.

If you suffer from frequent or lengthy losses of power, you may want to consider having a battery back-up system installed. In fact, if some appliances must absolutely run all the time, you should definitely consider a battery back-up system.

Battery backup system considerations

There are two types of battery back-up system for solar PV systems.

DC – coupled systems
These are better known as true 'off-grid' systems. The solar panels are connected via a solar controller to a battery bank, to which an AC inverter is connected. These systems have

92-94% efficiency, as opposed to grid connect inverter efficiencies of 95-97%. Batteries are either lead acid or lithium Ion: the latter, though much more expensive, are smaller, longer-lasting, and more efficient, and becoming more popular as a result.

AC – coupled systems
These are generally considered if there is one or more renewable energy source in a single system, such as a combination of wind, hydro and solar. Combining all of these sources of power in order to take advantage of them during a loss of grid supply is best done via an AC coupling. The trick in an AC coupled system is to keep the grid-connected inverter operating during a power cut, with the battery back-up source acting as a utility power supply so that the inverter operates in the normal way. In fact, AC coupling is pretty straightforward if a 230V battery back-up inverter is used in conjunction with a 230V grid-connected inverter.

Battery sizing is much more critical than in a typical grid-connected system, and detailed calculations are essential. Of course, the provision of running loads on a grid-connected renewable generation system is simple. If the renewables cannot provide the required power, the extra comes straight from the grid. But if you size a battery back-up system and it is too small to meet your load requirements, it will probably run out of power during a power cut – which rather defeats the object!

Renewable installer Bright Green Energy firmly believes that 100% of the electricity generated by your solar PV system should be consumed and not exported. Exporting unused electricity really does not make financial sense whatsoever. If you are out at work all day, and you discover that you are exporting more electricity that you would like, you can install a device that will switch on and off your immersion heater, dependent on the power produced by the solar panels. This approach is simple and easy to achieve, regardless of whether or not you receive the FIT payments.

However, what if you simply cannot achieve 100% consumption and are still exporting? The simple

AVOID 'ISLANDING'!

You might think it impossible to use self-generated power in the house during a power cut, not least because it flouts the UK's G83/1 rule that the inverter must disconnect itself during a power cut. This is intended to prevent the oddly named 'islanding,' whereby a property returns energy into a grid system that is supposedly shut down, with potential safety implications for line workers and other households. For that reason, all approved PV and turbine inverters have to shut themselves down when an absence of mains power is detected. But there are safe ways around this – see ATS systems – page 20.

answer is to store the electricity in deep cycle batteries ... and use it when you want!

On-grid with battery back-up

This particular system is based upon what Bright Green Energy's Alex Kennedy describes here. Alex explains: "In most cases, if you already have a solar PV system installed, you can either 1) replace the existing inverter with an inverter that has in-built battery support, or 2) install an additional inverter/charger alongside the existing grid-connected inverter. It is true to say that an inverter/charger's 'secret' is that it can take advantage of the fact that the AC OUT connections (terminals) of most battery back-up inverters can allow power flow in either direction. Normally, these terminals are considered as supply-only connections. In the ideal situation, the battery back-up inverter will have a contactor or electronic switch that will directly connect the AC OUT to AC IN connections when the grid supply is available. This means that, when the grid is available, it functions as the synchronisation signal for the grid inverter. When the grid goes down, the relay/contactor will open and the AC OUT connection functions as an AC voltage source that has all the characteristics of the grid, while the AC IN connection is disconnected from the grid. Of course, there are limits with this approach as battery power and inverter rating may affect pass-through capacity when the grid is down."

It seems clear that this is an approach that should only be followed by those with sufficient technical knowledge, and, in the UK, MCS accreditation – which Bright Green Energy has.

Alex Kennedy goes on to say that, in the opinion of Bright Green Energy, the backed-up loads must be separated from non-backed-up loads and placed in a separate distribution board or separated out in a split, dual RCD board. Alex continues: "Determination of the load requirements is the key to success. Careful analysis of the loads to be supported by batteries in the event of loss of grid supply, or power to be consumed at night, requires careful analysis. If, for example, it is required to run a microwave oven, small laptop, fridge/freezer and a few, preferably LED, lights, then these are considered as standby loads, and should be separated out from the normal consumer unit. In the event of loss of power, the standby loads will be run directly from the batteries."

BATTERY STORAGE
This can be added to renewable energy systems in one of three main ways –

1) Manual extension
The cheapest way of adding battery support to an existing solar PV system is to use a simple battery charger, inverter (or inverter charger), manual changeover switch and new separate fuseboard, if required. The decision to install a separate fuseboard should be based on a sensible approach to the home wiring already in place. A qualified electrician is required to install the system.

This simple and easy to install system means that the batteries can be charged (for instance) during the day from solar PV system, and used at night. This system is designed for low power use and is operated manually. Operation depends upon switching over to battery power when a renewable energy system is not supplying power.

Battery size (storage capacity) depends on load demand, but a minimum of 2 x 220Ah AGM batteries should be considered of practical use, though 4 x 220Ah is much more practical. Note: these small battery systems will be able to run reasonable loads but will not power an entire house. Load analysis must be carried out previously as batteries are affected by how often and how deep they are cycled.

2) Inverter extension
Either through the addition of new and separate battery inverter system, such as SMA's Sunny Backup, or via the addition of a battery inverter system – Victron being a good example – and a solar switch.

Some inverter manufacturers such as SMA already produce a battery back-up inverter that is installed as an add-on to the solar PV system. The Sunny Backup (not, at the time of writing, certified for use in the UK) will automatically switch to battery back-up with no effect on the efficiency of the solar PV system, and the batteries can be utilised to power appliances during night-time hours.

The Sunny Backup should be used in conjunction with the SMA Meter Box. This ensures that maximum self-consumption takes place during the night, and results in greater independence from the electricity grid by reducing the amount of power drawn from the grid. The result? Increased efficiency and effectiveness of self-consumption systems, as well as increased efficiency of the battery. Capacities from 2.2kW to 100kW available

3) Inverter with in-built battery support
Some inverter manufacturers offer built-in battery support. These are ideal for new solar PV systems as these inverters support direct battery connection. Alternatively, they can be installed as a replacement inverter for an existing solar PV inverter.

A good example of an inverter with built-in battery support is the Nedap PowerRouter. This is a modular, smart inverter with an integrated battery manager, which can be used for on- or off-grid applications. The inverter is designed to maximise self-use: in other words, the available solar electricity can be 'routed' directly to the loads or stored in batteries for later use. The Nedap approach allows the batteries to be correctly cycled, thus extending their life. Just like the Sunny Portal, the PowerRouter can be connected directly to the internet via a web portal. The PowerRouter is available in 3.0kW, 3.7kW and 5.0kW versions.

Chapter 4
Solar thermal – domestic water heating

PART 1: OVERVIEW

Solar water heating (or 'solar thermal') systems use panels fitted (usually) to your roof to heat domestic hot water. It's generally estimated that solar water heating can meet over half of your hot water needs* – although, of course, this depends on which part of the world you live in, the orientation of your roof, and the number of panels you install, as well as your domestic consumption needs.

*The UK's Energy Saving Trust, Field Study 2011 states: 'From the properties we trialled, well-installed and properly used systems provided around 60 per cent of a household's hot water.'

HOW IT WORKS

As the sun's rays connect with a solar thermal collector, an electro-magnetic stimulation of particles in the collector's absorber tube or plate occurs, which causes light energy to be transposed into heat energy. The heat is then transferred, either directly or indirectly, to the hot water system of the residence. Solar thermal systems produce hot water without creating any carbon dioxide, or any other greenhouse gas, and create no pollution. They produce energy without

SUPPLEMENTARY HEATING

In almost all circumstances – and certainly in cooler parts of the world – it will be essential to have one or more supplementary systems to 'top-up' the hot water supply when required, and to supply heat when the sun isn't shining. If the supplementary heating uses gas, oil or electricity, The Energy Saving Trust has proved that getting the timing right will be crucial to saving money and making the most of solar heating.

The EST says: 'The timing of the back-up heating system and hot water use has been found to have an impact on how much energy a solar water heating system can provide. Systems which provided more energy were found to time back-up heating to finish just before the period of higher hot water use, and commonly (although not exclusively) at the end of the day. Subsequent use of hot water from the cylinder in the evening or in the morning lets cold water into the base of the cylinder, which provides a volume of cool water for the solar collector to heat up the

next day. Many of the solar water heating systems where back-up heating was found to be adding heat to the cylinder after hot water use had ceased, provided less energy overall, as there was less or no cool water for the system to heat.'

CENTRAL HEATING

Rarely, if ever, is enough hot water generated to contribute to a central heating system, although, since around 6% of the total energy use in the UK is accounted for by domestic water heating, a lot of money could be saved and CO_2 emissions reduced if everyone who was able to used solar water heating. However, in a very small number of installations – and if designed and sized appropriately – a solar thermal system could provide a limited contribution to space heating, especially if there is an under-floor heating system working at lower temperatures than traditional wall-mounted radiators.

using scarce fossil fuels and are cost-effective. The light can strike the collectors in the form of direct radiation (direct sunshine) or diffusion-radiation (the kind of light that you get on a cloudy day).

OPTIONS EXPLAINED

Much of this information derives from an invaluable source produced by ATAG Heating Group, whose core business is the manufacture and marketing of high quality condensing heating boilers, thermal solar and flat-plate panels (their thermal solar branded panels are used in more than 55 countries). Although this chapter covers flat-plate and evacuated tube systems, ATAG is not an installation company, but supplies installers and associated businesses. Other illustrations and information are courtesy Eco2Solar Ltd.

HOW TO CHOOSE A DOMESTIC SOLAR WATER HEATING SYSTEM
1. Evacuated tube collectors

Evacuated tubes are often described as being more efficient than flat-plate versions, and can take up less space. However, there is some evidence that there may not be much difference between the two, and some installers fit a slightly larger area of flat-plate than would be used in an evacuated tube installation to make up for any possible shortfall. Here, our Eco2Solar evacuated tube heat collectors are beginning to create a small but useful amount of hot water early on a sunny but very frosty winter's morning. 'Shadowing' is far less detrimental than with solar PV panels.

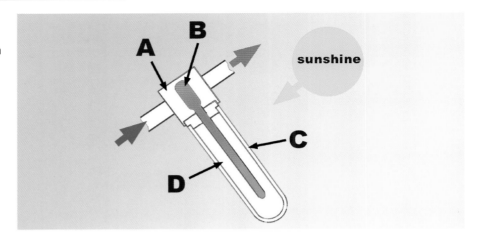

A vacuum between glass tubing provides extremely efficient insulation, and evacuated tubes can reach very high temperatures. There are two different categories of evacuated tube –
• direct flow • heat pipe
– which are similar in appearance but work in different ways. Heat pipe evacuated tubes can only be installed vertically, while direct flow tubes can be installed vertically or horizontally.

In the case of vacuum tubes, blackbody (defined as 'a hypothetically perfect absorber and radiator of energy, with no reflecting power') is enclosed within a glass tube (D), which is then made into (or surrounded by) a vacuum (C) which, since light can travel through a vacuum but heat cannot, prevents almost all heat loss. The glass tube is connected to a metal condenser (B) and the heat is directed to the condenser. This heats water in the manifold (A) which is transferred to the heat store. The condenser is sealed and contains a volatile (and safe) liquid which turns to gas when heated, and rises to the top of the tube, which.fits into a manifold. The gas in the tube heats the water in the manifold and, as it gives up its heat, returns to liquid and runs back down the tube ready to be reheated.

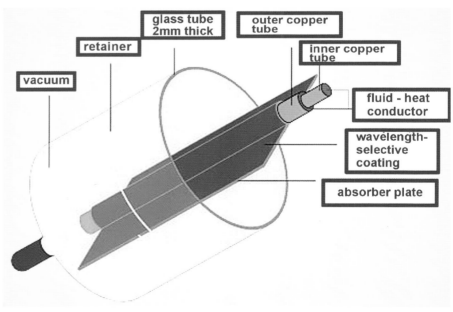

There are varlations on this basic theme, such as the 'Ecotube' shown here and fitted by Eco2Solar, who claim that the Ecotube's performance exceeds that of competing evacuated tube systems, including having much greater longevity. I certainly have experience of inferior tubes failing after a very short time.

Flat plate panels

The light falls on the absorber plate of the collector, which is coated with a substance that ensures maximum absorption of light. Copper tubes behind the absorber plate collect and transfer heat energy via heat transfer fluid. From there, the heat is drawn as required into a hot water system.

Some flat panels are covered with acrylic domes. Generally, experience in Germany has shown that regardless of the collector, many acrylic coverings have developed cracks at the points where they are attached to the collector frame. It is believed that these cracks occur as a result of the acrylic becoming brittle after long use while stress results from major changes in temperature. These hairline cracks do not generally impede collector performance in the short term. Glass does fracture, although long-term tests in Germany show that this is a very rare phenomenon.

Combined panels: at the time of writing, these have not yet been developed but the idea is interesting! If you haven't got sufficient roof space for all the solar panels you want, or if you want to reduce their visual impact, combined photovoltaic and thermal arrays would solve the problem. Now, if only they can be made to work ...

Drain-back systems

Because of the Dutch Water Board's preference for not using any chemicals, the drain-back was adopted in the Netherlands. The system is configured in the usual way, except that the water – in some cases, the domestic hot water – flows thorough the panel. However, when the pump switches off (because the solar is not heating the water), the water inside the solar panel will drain back into a small bottle. The system is thus protected against damage from boiling and freezing.

Pumping domestic hot water through the panel can lead to scale build-up and a steady decline in performance. No drain-back system can be truly sealed or pressurised. With drain-back it's imperative that there is a continuous fall-back to the reservoir. If the pipework has not been very carefully designed and constructed, the installation will not work, either due to air pockets or frost damage. Because of this, some installers of indirect drain-back systems use glycol instead of water.

- As sealed systems get hotter than drain-back systems, the plumbing must be of higher quality and hard-soldered (brazed) and not soft-soldered.
- Freezing is the biggest potential problem, but edible glycol prevents this, with a gel point of -30°C. Drain-back systems have to use more powerful, noisier, less efficient pumps, such as gear pumps, and cannot work below freezing point.

Advantages of drain-back systems

- Simple design and faster installations with lower grade pipework required
- System not under pressure
- No chemicals in the system (sometimes)
- Water can carry a little more energy than glycol

Disadvantages of drain-back systems

- Often not suitable to adapt to existing domestic system
- The cylinder must be located such that the pipe can fall correctly
- The pipework must have a continuous fall to avoid air locks
- Failure can result in costly damage to the solar panel
- The control system has a long start-up time as the water has to travel to the panel before energy can be absorbed
- Energy can be lost back to the atmosphere
- Requires water level to be regularly checked
- If the system boils the liquid will be lost
- Difficulty ensuring reliable drain-back and refilling when several collectors used

Plastic thermal panels

There are systems on the market which comprise a plastic thermal panel and a small PV cell driving a small dc motor. In such a system, the domestic hot water is pumped through the panel and then into the cylinder. The system requires very little plumbing. It does not have freeze protection because the plastic pipes it uses are designed to freeze without damage being caused. These systems do not provide a good solar fraction (energy provided divided by energy required) – rarely do they achieve more than 50%, and are incapable of providing very hot water, although they can make a valuable contribution to a domestic hot water system.

Although they have the advantage of saving electricity and being able to function in a power cut, the savings of not having a controller and pump hardwired to the mains amount to only 9 kWh per year (at the time of writing), although, of course, that could change if energy prices soar.

Unglazed solar collector

Although this type of collector absorbs a slightly higher degree of the sun's energy, because it is not insulated a high portion of the heat is lost, particularly on windy days. This system is primarily used for heating swimming pools, where pool water circulates directly through it.

WHICH IS BEST?

The following are the admirably fair and well-balanced views of ATAG, supplier of flat-panel systems.

Reasons to install vacuum tubes –
- They are quite easy to fit as the manifold can be mounted on the roof and the tubes carried up to the roof by two people. Flat-plate panels can be heavy. ATAG panel weight is

around 40 kilos (88lb), and requires two people to lift them onto the roof

- Tubes perform slightly better in relation to their size: generally, a vacuum installation requires around 10-12% less roof space than an equivalent flat-plate system
- If an individual tube fails it can be replaced; the entire manifold of tubes do not require replacement. There is a relatively much higher failure rate of individual tubes (compared to the failure rate of well-engineered individual panels). A tube failure is indicated either when fogging is apparent or frost is not visible on a particular tube when it is visible on other tubes in the same manifold
- Virtually all flat-plate panels are vented to allow the escape of condensation and expansion and contraction of air as the absorber surface heats up. Venting means that dirt can adhere to the surface, marginally interfering with the efficiency of the absorber plate. At locations where unusual conditions can occur, such as when close to the sea, or in cities where a thermal inversion effect takes place (Mexico City, Santiago, and Bath, for example), pollution can enter the panel and adhere to the absorber surface, substantially shortening its useful efficient working life. As the ATAG 1450 is evacuated to 100 Pa, the absorber surface is enclosed within a sealed casing, thereby preventing ingress of salt or other chemical
- The absorber surface is selectively coated with a thin layer of a substance based on colloidal nickel pigmented alumina. As the surface expands and contracts some wear and tear occurs, and there is a risk that, in time, the surface will become partly detached from its coating, meaning that although the surface will still be able to absorb heat well, it will not do so as well as previously. ATAG believes that any detachment is extremely unlikely, and in any event will not occur for at least 25 years but it feels that the possibility should be pointed out

Reasons to install flat-plate panels

- Tubes are prone to overheating (which reduces system life) because

it is difficult to design tube systems in such a way that avoids this. Some tube systems try to overcome heating problems by incorporating automatic valves in the manifold. Flat-plate panels rarely, if ever, suffer from overheating, and therefore tend to last longer than tube systems

- Well-engineered panels are much more physically robust than tubes
- The stresses caused by the expansion and contraction of the glass in tube systems (the coefficient of expansion of glass and metal are not identical) can result in stress where the glass is joined to the condenser, and sometimes stress fractures are caused, which means that the vacuum fails. The whole of the flat plate panel system is made of metal; glass covers the absorber plate but rests on a special washer which allows for the differential in expansion and contraction rates
- The vacuum seal, located as it is in glass tubing where this meets the manifold, is actually located upon the hottest part of the collector, causing stress on the seal
- The system of holding the tubes in a manifold and securing them to a roof means that, in windy conditions, minor tube movement can create glass fractures, which lead to vacuum failure. Panels do not suffer from this problem
- In snowy conditions (not a major consideration in most parts of the UK) snow tends to remain in the gaps between the tubes, reducing efficiency, whereas it tends to slide off panels much sooner
- Panels can be roof integrated, and are actually generally cheaper to install in new-build situations. It is not possible to integrate tubes into the roof
- Installations with vacuum tubes usually require more service calls than installations with panels, owing to the more fragile tube construction. The aesthetics of tubes and panels are a matter of personal taste. Some people like an array of futuristic-looking tubes on the roof, whereas others prefer that panels can be roof-integrated

The German market initially favoured tubes, but today, flat-plate panels have a massive 82% of the market.

How many panels?

As a very rough rule of thumb, one square metre of collector area on the roof equates to each person in the household. In addition, each metre of panel area will need a hot water tank with a volume of between 30 and 60 litres. Also bear in mind that if you use the slightly less efficient flat-plate solar water heating panels, you'll need to cover a slightly larger area than if you use evacuated tube collectors.

Positioning

IMPORTANT NOTE: In the southern hemisphere, all north-south references are reversed.

It's also important to choose the correct location for your solar panels, and you'll need to consider these main factors –

- shade from trees or surrounding buildings (at all times of the day!)
- the pitch of the roof (which has an impact on the amount of energy the collector can absorb)
- the unshaded area of south-facing (in the northern hemisphere) roof available
- position of pipe runs inside the house
- future access for servicing and maintenance
- the amount of hot water you are likely to use
- the type of water heating system already fitted
- the amount of money you have to spend
- south-facing roofs: collectors should ideally be mounted at an angle of between 30 and 60 degrees, and not shaded by overhanging trees, buildings or structures. Good results can be obtained by positioning the panels facing anywhere in an arc between south west and south east, though due south-facing provides optimum performance
- east/west facing roofs: good results can be obtained by splitting the collector array on east/west elevations, but this should only be done when it is not possible to have a mainly southerly position for the collectors. If installing a split east/ west system, the number of panels should be doubled:, a two-panel system requires two panels facing west and another two panels facing

east; each set of panels will behave as an independent system. However, where roof pitches vary or there is a lack of space, good results may be achieved by installing an additional panel on either the east or west orientation
• north-facing roofs: not generally recommended in the northern hemisphere

FLAT-ROOF SOLAR COLLECTORS
See also Chapter 6: *Flat-roof and free-standing frames.*

At north European and northern American latitudes, the orientation of a solar collector should always be as southerly as possible (for optimal performance), with, ideally, a 35-degree pitch. If the collector has to be laid horizontally (due to being in a conservation area, for example), a direct evacuated tube, oriented in an east/west position and rotated to 35 degrees, will be the most efficient.

Solar collectors can also be mounted on an A-frame on a flat roof, but the frame will require either significant weighing down (the roof needs to be strong enough to accommodate this), or being tied into the wall structure to prevent wind uplift. A structural engineer can advise on this.

Cleaning
A problem with some flat-plate collectors is that dirt can accumulate externally, and long-term tests show that, without cleaning, at various locations optical efficiency is reduced by between 0.3% and 1% after 8 to 16 years. With acrylic domes it was found that thorough cleaning could not be achieved because the dirt tends to be 'burnt on,' and this reduces optical efficiency more significantly. With glass collectors the rain washes dirt away easily, and cleaning is not recommended unless the collector is installed in an exceptionally dry, polluted environment which experiences high dust conditions. The glass that covers vacuum tubes also allows dirt to be washed away easily, but in some locations moss accumulates on the underside of tubes, which tends to affect the radiation transmission through the side of the tubes, and in some tests was observed to reduce efficiency by 15%

(when compared with new). Many flat-plate panels collect dirt on the interior of the glass, which usually looks worse than it actually is: in tests that compared the efficiency of dirty and clean glass in glazed collectors, a drop in efficiency of just 0.4% was recorded.

ADAPTING EXISTING DOMESTIC HOT WATER SYSTEMS
Your existing system probably won't have suitable components to do this, such as an appropriate type and size of hot water cylinder, and you will always need extra controls and pipework. However, it is possible to add solar thermal panels by adapting most existing hot water systems, in which case, you'll need an additional cylinder

for pre-heated water, or preferably change your existing cylinder for a twin- (or more) coil cylinder – see later in this chapter.

Combi boilers
It's usually not practicable to add solar water heating to a combi boiler system. These are designed to take mains pressure cold water, whereas solar water heating systems supply warm water at low pressure. A few combi boilers will accept pre-heated water – check with manufacturers.

PLANNING PERMISSION
In the UK, planning permission is not usually required to install solar heating, unless yours is a listed building, or is in

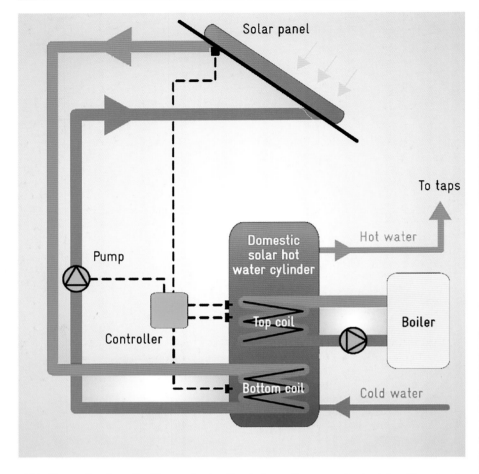

This illustration from the Energy Saving Trust shows the general principle of how heat is transferred from the collector on the roof to the hot water cylinder, using a self-contained volume of fluid that is not mixed with the domestic water.

Earlier, reference was made to the temperatures and output figures. A controller should be connected into the system which provides all of this information at the touch of a button.

This location is the pitched roof of a building.

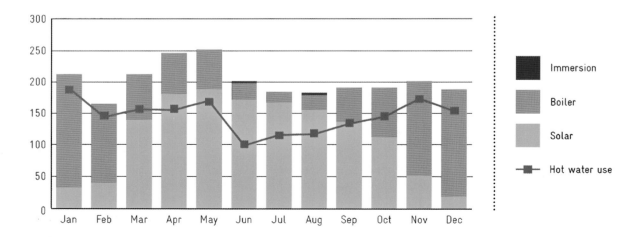

This Energy Saving Trust graph shows a typical energy profile for a well-installed and properly run solar water heating system. The solar energy input to the hot water cylinder is at maximum in summer, with back-up heating providing more energy in the winter months.

This Eco2Solar installation, comprising both PV and solar thermal panels, is on a large, flat roof.

They even installed one on a kitchen wall! It could be argued that maximum efficiency could never be achieved, but this outcome was found to be quite acceptable, and making a waterproof entry into and exit from the house for pipes was relatively simple.

This Eco2Solar installation shows how a garden pergola has been adapted to accommodate solar thermal panels, which could be an ideal way to position the solar thermal system near the swimming pool. Ubbink (and others) makes free-standing solar collectors (inset) especially for this purpose. Typically required are (very approximate figures) –

With or without cover (indoor pools can expect a slightly lower heat loss)	Flat-plate	Evacuated tube
Outdoor pool with cover	1 flat-plate collector per 5 square metres of pool surface area	6 tubes per 4 square metres of pool surface area
Outdoor pool, no cover	1 flat-plate collector per 2 square metres of pool surface area	6 tubes per 1.5 square metres of pool surface area.

Although this book mainly deals with renewables for the home, it's worth pointing out that large savings can be made by using solar energy to help heat swimming pools.

A downside of free-standing solar thermal installations is that a relatively long pipe run, with greater heat losses, will be required.

Ideally, the exposed pipe run should be as short as possible, consistent with being able to achieve the required entry and exit points.

Here, Eco2Solar has constructed lead entry/exit points, and built them into the structure of the roof. It's the only way of ensuring a completely watertight roof.

The other major consideration when an installer is debating where to penetrate a roof, is to ensure that there is good access from the inside for connecting and running the required pipework.

a conservation areas or world heritage site. Contact the Energy Saving Trust (EST), which has lots more information about planning permission, or your local planning authority. Other countiies may have similar planning regulations.

INSTALLER CHOICE
It is strongly recommended, as with all the technologies described in this book, that a good deal of research is done before appointing an installer, and that, in the UK, you use only MCS-certified installers and products. Watch out for the dodgy sales tactics employed by low-quality companies, which are prone to exaggerating the financial savings that can be made.

INSTALLATION OVERVIEW FROM ECO-NOMICAL
'There are many ways to install a domestic solar hot water system, which calls for similar skills to those required to fit a central heating system, with the added ingredient of having to gain access to the roof, just to spice things up! Keep in mind, though, the big difference, which is that the solar circuit can reach much higher temperatures than regular domestic plumbing; possibly well in excess of 100°C. Because of this, it is important

that all components should be able to withstand these high temperatures. Also, position the components in such a way as to minimise their exposure. Place pumps, controls, etc, on the return leg of the circuit, so that the solar fluid has been cooled by the water in the hot water cylinder. Even high temperature solar expansion vessels are only rated for 110°C, so place these on an uninsulated branch a little way from the solar circuit. The use of a pump station makes for a neat installation, and fewer individual components to worry about. Make sure that the solar loop is well insulated with high temperature insulation. A poorly-insulated solar circuit can cause a significantly greater loss of efficiency than sub-optimal collector orientation.'

PART 2: WHAT YOU NEED
The vast majority of people will commission an installer to design and specify their system for them. Here's

Insulation: having gone to a great deal of effort to capture the energy, equal care must be taken during installation to minimise heat energy losses by conduction through the system's pipework. Using the correct insulation is critical, since it must endure a temperature of more than 170°C near the collector. Normal central heating insulation will melt and/or become brittle at these temperatures. Where the insulation is outdoors, it must also be resistant to air pollution, UV radiation, and pests such as mice and birds that may eat it. In some countries insect activity necessitates insulation be 'armoured.' Insulation should be externally sealed to prevent it carrying moisture. If insulation is sheathed, care must be taken to seal lap joints, ideally with aluminium foil, allowing for proper overlap. This is likely to ensure a life of at least 25 years.

an overview of the kind of equipment that will be used in installing a solar thermal system, apart from the panels that are dealt with in Part 1. Again, much of this material is based around information supplied by ATAG and associated specialist companies.

Pipework

The hydraulic system's pipework should consist of flexible, annular pipe, made from stainless steel, and straight seam welded. It should bend to a radius of 100mm (4in) and have a maximum operating pressure of 10 Bar and a burst pressure of 100 Bar: 3mm thick; 25 meter lengths; factory insulated with 25mm closed-cell insulation.

Copper pipework: Soft solder should not be used in solar installations because of the very high

Annular 'Twinway' corrugated stainless steel pipe has flexibility, and is pre-insulated to save time and money on installations. It invariably comes with clamped fittings to form a BSP female nut connection on each pipe end. It's also available with an integral temperature sensor cable.

temperatures that can be experienced – well in excess of 100°C. While soft solder has a melting point of around 200°C, ATAG collectors, for instance, have a stagnation temperature of 170°C. So, when the pump starts in the exchange system, bursts of temperature well in excess of 200°C are possible (albeit for short periods), which might melt or weaken the joints. (Fernox a good quality soft solder - has a melting point of 227-228°C, but even this is not quite high enough.) Hard solder has a much higher melting point and will result in no weeps or leaks. The only possibility of a chemical reaction between the solder and the glycol arises if a solder has a high zinc

content, so no zinc, or very low zinc content hard solder must be used.

Copper jointing: ATAG recommends the use of press tools, powered by a rechargeable battery (see Chapter 5, *Biomass central heating boilers*). Copper jointing creates extremely sound, pressure-proof joints without soldering or brazing: particularly important as the glycol solution has a much lower surface tension than water, and will find cracks and crevices in pipework and jointing that water will not.

ATAG does not use mono propylene glycol, but polypropylene

Evidence from hundreds of thousands of installations leads ATAG to conclude that the glycol chosen is far more important than many installers realise. Factory-mixed solutions should be used, because, when glycol is mixed onsite, it tends not to be mixed with distilled water, so introducing some impurities into the solution, and it's rarely mixed thoroughly, allowing crystals to form under the high temperature differentials experienced. These formations, of course, weaken the anti-freeze properties of the glycol, with freeze-damage to the heat exchange pipes and pipes in the collectors the result. ATAG solutions protect down to -30°C. (You must check the protection level of whatever brand you or your installer chooses.) Over a 10- or 15-year period, all glycol loses its anti-freeze properties to some extent. After 10 years, ATAG fluid will still protect down to -11°C.

glycol, which is much more suitable because it is approved for use in the food industry: it is safe and edible. Under no circumstances should a ethylene-based glycol be used in solar systems.

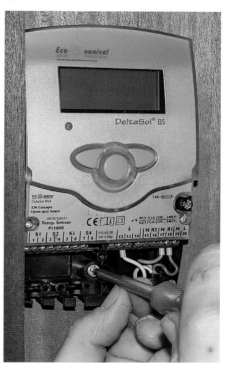

The digital controller receives information from the sensors and controls the flow of the hydraulic system, according to need and energy availability. It is programmable, but ATAG's version is not intended to be user-programmable, although others, such as the unit we fitted, have some user settings included. Electrical installations must always be carried out by qualified personnel.

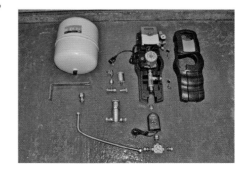

The pumping station is made up of various components and, in the better systems, each component is 'modular' so that it can be separately replaced.

User safety: there are many safety controls built into every ATAG system, as there should be in all systems.

- Overheating: ATAG panels have a stagnation temperature of 170°C; the system has as a minimum an 18 litre expansion vessel, and a pressure relief valve set to 6 Bar. The system is pressurised to 3 Bar but tested independently to still remain viable at 10 Bar
- When glycol vaporises in the panels (which occurs at around 140°C) the vapour volume is pushed into the expansion vessel. Even on the hottest days the pressure relief valve should not need to activate
- For temperatures in excess of 180°C, there is a solar pressure relief valve. All other components used in the system are temperature and pressure rated well over their anticipated operating temperatures and pressure on the system. ATAG recommends, if there is room, that the discharge from the expansion relief valve is run to a container within the vicinity of the pump station. This container should be twice the size of the volume of glycol above the valve
- ATAG controllers will shut off the pump when the desired predetermined temperature is reached. Heat exchange piping must be insulated with solar quality insulation, so that there is no exposed pipework inside the house
- If an unvented cylinder is used, ATAG will install an additional, two-port valve to shut off flow from the solar thermal system, if overheating should occur. The overheat thermostat is wired via the solar and heating controls, so that all solar thermal operations cease, providing an easy visual indicator to the homeowner that something is wrong. It must be manually reset as an additional safety measure
- The solar thermal system is sealed, so the heat transfer fluid does not at any time come into contact with the domestic hot water

Precisely the same heat exchange operations take place within the cylinder, whether or not it is fitted with a solar coil. The life of the cylinder depends largely on the quality of the water flowing through it. However, it is important that, in pressurised systems, the solar coil inside the cylinder can withstand the pressures applied. System designers should check manufacturer specification. This cylinder, produced by solar thermal specialist Newark Cylinders, is designed to be suitable for the use specified.

This earlier solar thermal system has reflectors that sit behind the tubes, but research has shown that they have little or no effect.

You don't *need* this – but it's something to consider! The author is in the process of setting up a small solar PV panel, 12V battery and inverter to power the solar thermal water pump. It saves a small amount of electricity/CO_2, and means that, in the event of a power cut, the system will keep working. In high summer, pump failure will lead to the system overheating and discharging its glycol as a safety shut-down measure – a fairly expensive and time-consuming business to put right.

The tubes or panels and other equipment might need to be delivered before the installers arrive, in which case you'll need a secure space in which to store them.

PART 3: ON THE ROOF
Introduction

This is the section of the installation that poses the greatest physical risk, because working on the roof is inherently dangerous, for obvious reasons. We strongly recommend that only professional installers, or those with sufficient professional training, carry out this work. Although the vast majority of solar thermal installations are fitted to roofs, some are fitted at ground level or on flat roofs.

Important notes

ATAG, whose staff assisted with the information here on flat-plate systems, points out that its solar thermal systems are not intended to be installed by do-it-yourself, unskilled or untrained installers, and states: "Installers usually have many plumbing qualifications, and are trained to use scaffolding and proper up-to-date safety equipment, and are aware of the dangers of working at heights. Fitting panels onto the roof is something that should only be undertaken by qualified professionals." This part of the work is described here for information only.

Eco-Nomical specialises in supplying components to both the trade and those competent to fit them. The instructions for evacuated tube systems shown here are based on text supplied by Simon Watts of Eco-Nomical; most of the pictures were taken during an installation of the company's components.

Eco2Solar is a UK MCS accredited installer of evacuated tube systems, and a large number of pictures in this book are of its work. Founder and Managing Director Paul Hutchens established Eco2Solar in 2007 to address the threat of climate change, by reducing CO_2 emissions and fuel shortages through the application of solar energy. "These days," he says, "following the introduction of government financial incentives, our clients are generally more interested in making money from their solar energy system!" His view, like that of almost everyone involved in this book, has much deeper ethical bias.

If planning to install solar thermal in accordance with a government-run financial incentive scheme, you may need to use accredited installers in

order to qualify. For instance, in the UK, the Microgeneration Certification Scheme (MCS) is the recognised quality assurance control, supported by the Department of Energy and Climate Change. MCS certifies microgeneration technologies used to produce electricity and heat from renewable sources. MCS-approved installation is (at the time of writing) a requirement in order to quality for the UK government's financial incentives, which include the feed-in tariff and the renewable heat incentive.

Professional installers must remember to –
- always use proper scaffolding and safety harnesses when working at height. This scaffolding was erected by professional scaffolders on behalf of Eco2Solar, prior to our second solar thermal installation
- always assess the risks before starting work
- take care when carrying the collectors to a roof. Carrying and manipulating heavy weights and large frames onto a roof is difficult, and can result in an accident
- always ensure there are a sufficient number of people for the work being undertaken
- always comply with all wiring and electrical instructions and regulations, including electrical bonding rules, when installing the pump station and digital controller
- be sure to have all necessary supports, harnesses and equipment for the job

SAFETY NOTES

When handling collectors, professional installers must remember that solar thermal panels have been designed to convert light into heat, and accordingly, parts of them will get very hot if left out in the sun, even for short periods of time, and severe burns can result from touching the affected parts. Some components can reach ultrahigh temperatures, in excess of 200°C, and should not be touched, whether or not the system is in operation. In sunny conditions, the installer must shield the absorber parts of the panels from direct sunlight until the installation is complete.

Collector arrays

Two large collectors may be plumbed together in series, but for an array of more than two (for commercial or other high volume use), it is recommended that they be connected in groups of two in parallel to reduce pumping losses, and also to reduce the risk of overheating in summer.

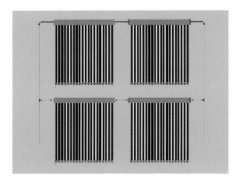

Four panels plumbed two-by-two in parallel.

EVACUATED TUBE – TYPICAL INSTALLATION DETAILS
Framework

Here's how to assemble an Eco-Nomical evacuated tube panel. Other types have similar installation processes, but always follow the supplier's instructions if assembling your own.

Eco-Nomical points out that it is possible for the inner, copper manifold to move slightly, relative to the manifold case, if the manifold has been dropped on its end while in its box. This results in the heat pipe pockets no longer being concentric with the tube holes in the manifold case, which makes it difficult to install the tubes. If this happens, with a piece of wood

gently tap the 22mm copper pipe at the relevant end until the manifold is correctly aligned with the tube holes once more.

Note that there are two types of tube fixing, and this first section deals with the clip type; if you have the screw cup-type fixing, see later.

Clip-type tube fixing

For ease of transport, Eco-Nomical evacuated tube solar water panels are supplied in kit form, as are almost all makes of panel. Subject to the agreement of your installation engineer, this might be an area where you could save money by carrying out construction of the framework on the ground. Check and identify the components of the panel. Justin Waters, assisted by his son, Matt, found a clear space in which he could work and laid out all the parts ready to hand, along with his toolbox.

Open the tube clips by releasing the screw on the jubilee type clip until the band separates from the screw block and then, push the loose end of each band through a housing on the base bar, one-by-one. For a neat appearance, push all the bands through in the same direction. Push the end of the band into its screw block, and tighten by a couple of turns. If the band is hard to start, rotate the screw backwards a little and try again.

Take the side bars and examine them. Near the bottom there are two upstanding lugs to which the base bar is bolted. Note that on some versions, the side bars are handed because of the additional holes for mounting the reflectors; the sides with the extra holes go on the inside. Fit the side bar to the base bar as shown, using one of the longer bolts to fix it. Do not over-tighten. Repeat for the opposite side bar.

REFLECTORS OR NOT REFLECTORS?

A 'mini-CPC' (Compound Parabolic Concentrator) is a highly reflective, weatherproof reflector fitted behind the collector tubes. A study jointly carried out by Chinese and British universities* measured performance from identical evacuated tube solar thermal systems, with and without reflectors. Surprisingly, perhaps, the results showed that: "... when attaining low temperature water, the evacuated tube solar water heater system without a mini-CPC reflector has higher thermal efficiencies than the system with a mini-CPC reflector. On the other hand, when attaining [very] high temperature water, the system with a mini-CPC reflector has higher thermal efficiencies."

However, in the mid-range of 65 to 75 degrees C, the maximum useful area for domestic hot water (DHW), there was virtually no difference between the two. Conclusion: for DHW, a system without a reflector actually provides better performance, and is also lighter and easier to lift onto the roof.

* *University of Science and Technology of China, and University of Nottingham, 2012.*

Now the manifold can be fitted to the assembled frame. Lay the manifold on its side on the ground, and fit the side bars to the captive bolts. Note that the holes in the side bars are elongated to allow some adjustment to suit the tubes. It's a good idea, says Eco-Nomical, to let the end of the side bars rest on the ground when the manifold is also lying on the ground.

Here's a useful tip from Justin: place the soft aluminium on some of the bubble wrap in which the kit has been supplied, to prevent damage while being worked on.

Locate the first reflector in the frame and insert and fasten a bolt. Notice that the reflectors have a bolt hole at each corner. The other three bolts are fastened ...

... and the procedure repeated for the other reflector.

Screw cup-type tube fitting
This is another design of tube clamp which uses plastic screw cups. The screw cup-type uses a different, extruded aluminium bottom rail which sits in a clip on the side rails.

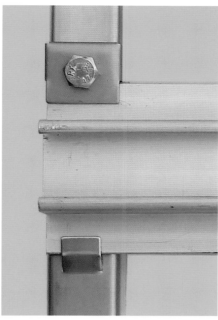

Fasten the rail via one bolt through the top hole. A rectangular, profiled washer holds the nut captive under the frame, and a metal clip clamps the rail in place.

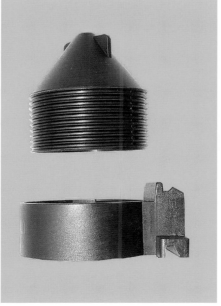

The Eco-Nomical tube supporting cup is in two pieces which screw together.

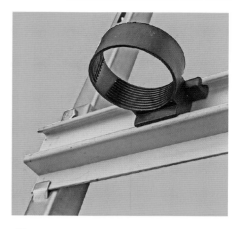

The ring snaps in to the mounting rail, or alternatively can be slid in from the end.

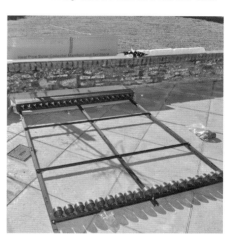

This is a slightly different version of frame, this time from Eco2Solar.

Justin and Matt started to manoeuvre the frame onto the roof a stage at a time, using ropes tied to the top of the frame to help secure it at the various stages of ascent.

ROOF WORK
Roof fixings

See page 47-on for information on roof fixings. It's essentially similar for both solar thermal and solar PV installations.

The installer will frequently require the use of roof ladders.

The type of frame used without reflectors is lighter, and can be appreciably easier to manoeuvre.

This was the Eco-Nomical framework in position with a covering on the lower part of the roof to protect it from damage when carrying out the work.

This is the seal and flashing that should be used where pipework passes through the roof.

Justin removed tiles from the relevant location ...

... and built in the flashing, having made a hole through the roofing felt (sarking) beneath.

This is the self-bleeding valve which Julian attached to the manifold with plumber's thread sealant, though he could have used PTFE tape.

The stainless steel hose was passed through the aperture in the roof ...

... and through the rubber seal ...

... which slides down the pipe and forms a seal on the collar.

Justin cut the pipe to length (see Part 4 for information on how to work with this type of pipe) ...

... slid the Armaflex insulation over it ...

... and secured the pipe with a compression fitting.

The rubber seal is held in place on the collar with a stainless steel jubilee clip.

When more than one collector panel is used, they should be joined together as described earlier.

The other end of the manifold is connected to a hose passing through the roof as close to the collector as possible. Do note that, sometimes, obstructions in the loft or underneath the roof space will require you to make a hole in a slightly less than ideal position, such as the one shown here. Note that a temperature sensor cable also passes through the roof at this point.

You must be careful not to damage the tubes while they are stored on the ground. If they are standing on end, leaning against a wall or similar, rest the end on a piece of cardboard or other soft material. It's better to store them in the packaging in which they came until you are ready to use them.

SAFETY NOTES!

- Do not install the collector tubes until the circuit plumbing is complete and the system filled with coolant. Placing the tubes in an unfilled manifold will quickly result in a very hot (and potentially dangerous) panel!
- In sunny conditions, the bulbs on the ends of the tubes can become hot enough to badly burn skin. Do not touch the bulbs for any reason after the tube has been in the sun. If in doubt, check the bulb temperature with an infrared digital thermometer before applying paste or touching it for any other reason.

With each tube assembly lying flat on the ground, thermal transfer paste is applied to the heat pipe bulb to ensure a good thermal contact between the heat pipe and the manifold. Spread the paste over the bulb in an even film.

You can also smear heat transfer paste onto the manifold seals to make inserting the tubes easier. Have a cloth handy to wipe your fingers!

Alternatively, use heatproof silicone lubricant spray where each tube will enter its seal once the tube is on the roof.

This is a different design of tube but the principle remains exactly the same. The person fitting the tubes on the roof will probably want to have an assistant on the ground applying the paste ...

... before carefully carrying the tubes up a ladder, one at a time, ready to pass to the person on the scaffolding.

Now, after checking that all of the tube rings are in place, each tube can be inserted. With the top end of the tube above the manifold, the bottom end is fed carefully through the first clip, or tube ring, from above, angling it to avoid the manifold. (The clip should be a very loose fit at this stage.) Then, the tube is slid back up again, while inserting the top of the tube into the manifold as shown here. Take care to locate the heat pipe bulb in the receptacle in the manifold. Twisting the tube helps to start it in the seal. Push home the tube.

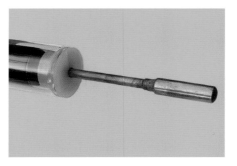

There's a slight difference in procedure with this type, where the bung in the top of the tube is different. Note that there is no copper collar above the bung, and the procedure for insertion into the manifold is slightly different. Pull the heat pipe out from the tube about 150mm (6in) before inserting. Hold the heat pipe tube below the bulb and push into the manifold pocket. Once the heat pipe is fully inserted, slide the glass tube into the manifold until the base of the tube protrudes about 15mm (0.6in) below the retaining ring, so that when the screw part of the ring is installed most of its thread is engaged.

Here you can see the tube on the left in the final stage of being inserted into the manifold before being pushed fully home.

Note that the glass tube should not be pushed up as far as it will go. Justin points out the correct positioning of the tube in its clip ...

... before screwing the bottom of the cup into place to retain the tube. Do not over-tighten: gentle hand-tightening is sufficient.

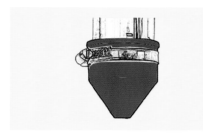

For frames with a clip-type fixing, slide the end cap over the tube and under the tube clip. Ensure that the tube is positioned so that the clip lies on the parallel part of the tube. Tighten the clip just sufficiently to prevent the tube from moving. Do not over-tighten, as this may cause the tube to fail. For a neat appearance, ensure that all caps are level, and all clips face in the same direction.

Installation manuals rarely mention this but it will be a minor miracle if your installer works on a tiled roof – especially one with old, oddly-shaped, hand-made tiles – without breaking any of them. Be sure that the correct replacements are available before work commences.

Interestingly, you can see the small amount of shade from an adjacent roof with the sun low on this winter's day. If this is something that occurs only at the beginning or end of the day at this time of the year, it's nothing to worry about. Unlike photovoltaic panels, solar thermal panels are not massively affected by a small amount of shade.

INSTALLATION DETAILS

For details of on-roof and in-roof mounting systems for solar thermal panels, see the relevant section in Chapter 6 where there is a great deal in common between solar PV and solar thermal installations, and where solar thermal systems are also covered. This is supplementary information relating to flat-plate installations.

Fitting of the panel and mounting framework will be entirely dependent on the individual manufacturer's instructions, which should be followed closely.

Changing pitch

Manufacturers such as ATAG have kits for use on low-pitched roofs which increase the panel pitch for a better inclination. This adds to the amount of wind stress that will be applied to the panel and the roof, and it is important that structural roof strength is considered, and that properly stressed, proprietary parts are used. In the case of the ATAG kits –

- 500mm increases the pitch by about 15°
- 750mm increases the pitch by about 21°
- 1000mm increases the pitch by about 27°

When the panels have been connected together and the pipework is complete, the system should be filled with drinking water, vented, checked for leaks, emptied, and re-filled with the correct water/glycol mix.

Important installation tips

- Use only solar-quality O-rings if using press fittings.
- Do not install the panels upside down – they will only work the right way up.
- Use support sleeves (pipes inserts).
- Never use plastic pipes for the thermal circuit.
- Join copper pipes by brazing or silver soldering, not with soft solder.
- The glycol is usually ready-mixed: do not add water or any other ingredient.
- Make sure the pressure vessel you use has a glycol-resistant membrane.
- Pressure test the panels before you connect up the circuit, especially if you are fitting them inside the roof.

Connecting panels together

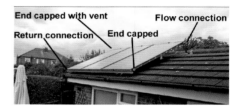

These panels are connected in parallel. The inlet (return) is always on the bottom left side and the outlet (flow) is on the top right side.

Apply a smear of silicone to the socket end of the connector and insert the O-ring.

Pull the panels together, making sure they are properly aligned, so that the connections are exactly opposite each other.

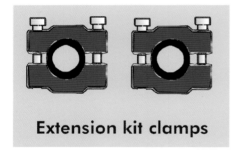

Extension kit clamps

When using this type of extension kit, apply a smear of the recommended sealant to the bolt thread and tighten with an Allen key.

Apply silicone to threads of bolt and clamp.

On the flow connection, attach the brass T-piece with air vent supplied using the same type of clamp and O-ring as before, and remembering to apply silicone to the threads.

PART 4: INDOORS

Use 15mm (0.6in) copper connected to the panels via compression fittings. Plastic pipe should not be used as it will not stand up to the prolonged high temperatures which the panels can produce. Soft solder joints should be avoided as, in situation where water, for some reason, is no longer being circulated through the panel and becomes stagnant, the temperature of the manifold could exceed the melting point of the solder.

These are typical of the components supplied by solar thermal companies, and comprise some of the hardware that will be fitted first.

On the left is the automatic air purging valve that will be fitted to the roof, at the highest point of the circuit. On the right is a Discal 551 de-aerator. This 'automatic air scoop' is a device which filters micro air bubbles from the solar fluid and ejects them via a high-temperature, automatic air vent within the device. It does not need to be fitted at the highest point of the circuit, and may obviate the need for the external air vent. However, the recommended, belt-and-braces approach is to fit them both. The Discal device operates within the temperature range of 0–110°C, so obviously must only be used on the return (cooler) side of the circuit.

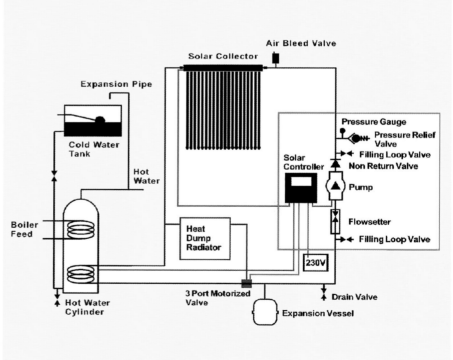

This typical solar plumbing layout produced by Eco-Nomical indicates the relationship between these components, though not, of course, their actual positions in the house.

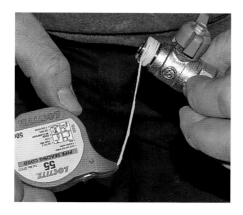

Ordinary white PTFE joint sealing tape is fine for use in high temperature applications: for example, at the pump station connections. Justin Walters of JW Solar Solutions demonstrates his preferred PTFE pipe sealant cord.

It makes sense to assemble any components that will have to go on the roof before taking them up there. (Inevitably, there will be a little more roof-work before the job is completed.)

FITTING THE HARDWARE
The installer should have made a plan of where all the various components are going to be fitted, and now is the time to start fitting them, before the pipework or wiring are connected.

PRESSURE VESSEL SIZING
The size of the pressure vessel is critical. These figures relate to ATAG 'Thermal Solar' flat panel collectors and, though they provide an idea of what is required, each installation should use a size of pressure vessel advised by the supplier of the collector panels being installed.

Number of ATAG solar collectors	Pressure vessel size in litres
2	12
3	18
4	24
5	32
6	36
7	42
8	48
9	56
10	60

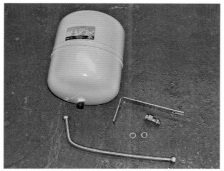

The pressure vessel (also known as expansion tank) must be fitted on the return side of the system ie: in the coolest available part of the system so that the internal membrane has a longer life span, and should be located in a position where it can be easily accessed. It should be fitted with an isolating valve for ease of replacement. However in ATAG's instructions to professional installers, it is recommended that the isolating valve should not be accessible without a tool so that an unqualified person cannot shut it off.

Zilmet, a manufacturer of these pressure vessels, recommends that –
- The vessel must be visibly inspected for pinholes in the metal body, and for connector leaks at least once a year (or as and when a drop in performance is noted from the system).
- The air pressure must be checked against the required pre-charge. Some pressure loss is to be

expected, and should be rectified to within a reasonable accuracy, but a significant drop in air pressure may signify that the vessel membrane is nearing the end of its life span, and may require replacement (which is possible without renewing the vessel itself).
- The air pressure may only be inspected when the vessel is either completely detached from the system or when the system itself is de-pressurised to atmospheric pressure.

This union fits to the bracket ...

... and Justin decided to fit the union and the top section of flexible pipe to the bracket before screwing it to a suitable roof tie beam. The fixings need to be strong because the potential weight of the pressure vessel will be approximately its capacity in litres expressed as kilogrammes, plus the weight of the vessel itself.

Justin screwed the adaptor with integral seals to the top of the vessel ...

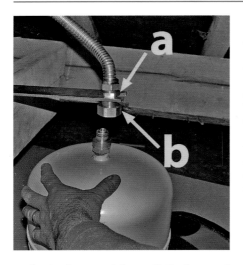

... the top brass nut, beneath the hose nut (a) was used to tighten the adapter to the bracket (see pic 8) and, on this model, the larger nut (b) is held with a spanner while the vessel is screwed hand-tight.

The pumping unit is provided with an insulating casing, which is fitted simultaneously with the unit. Justin passed a screw through a fixing hole and the insulation ...

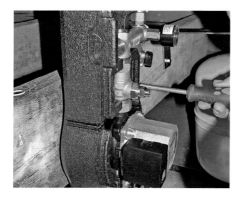

... before screwing the pumping station body onto its pre-planned location.

PUMPING SYSTEM

ATAG points out that the pump station and associated components must always be mounted at a level that is lower than the panels, so that no steam may penetrate the expansion tank in case of stagnation. Stagnation occurs when the solar panels have heat energy which is not needed by the system, and the panels stagnate at around 180° Celsius. If the expansion vessel is mounted at the same or higher level than the pump station, a thermal insulation loop must be installed. The pump station is not suitable for direct contact with swimming pool water.

NOTE: The solar hot water cylinder has (at least) two heat exchange coils within it. The upper coil is usually connected to a conventional boiler, to provide hot water at times when solar provision is inadequate, and/or another coil may be connected to another source, such as a log burner back boiler. The lower coil is connected to the solar circuit. The water in the lower half of the tank will be cooler than that in the upper half – warmer water rises to the top of the tank – which enhances the heat transfer from the solar circuit. A standard central heating pump is used to circulate the solar fluid, usually set to its lowest setting.

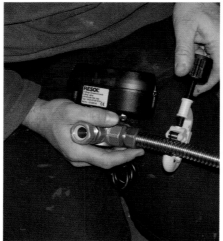

The Resol 3-way valve is fitted at the base of the pumping station and, again, Justin elected to fit pipe stubs to save having to tighten them onto a component already in situ. There are no hard-and-fast rules about this sort of thing; it's all a matter of what is most convenient.

Why, you might ask, is a fairly heavy, twin-panelled radiator being carried into the loft?

The ATAG flat panel system has a shut-down facility for when the domestic water has reached its designated highest temperature while this evacuated tube system dumps its excess, unwanted heat. As Eco-Nomical says: "Other than for small installations using unusually large hot water cylinders, if there is a possibility that the house may be vacant in summer, for instance during holidays, then a heat dump circuit should be employed to prevent the solar fluid from boiling. A heat dump works by bypassing the hot water cylinder heat exchanger, routing the solar fluid through a small radiator mounted in an attic or garage. When the hot water cylinder reaches its preset maximum temperature, a motorised valve diverts the flow through the radiator and dumps the heat. The radiator must be placed where the heat from it will not be a nuisance (since it will be working in summer), or a hazard (since it may well run hotter than a conventional radiator). A suitable controller is required to operate the valve. As always, the motorised valve must be suitable for high temperatures, and should be on the cooler, return leg of the circuit."

After measuring the correct locations of the radiator brackets, Justin fixed both brackets to the framework that he had earlier constructed in the loft ...

... then lifted each of the two radiators onto its brackets. Note that in order to prevent an excessive amount of heat reaching the bedroom beneath, the entire space between the bottom of the radiator and the ceiling was filled with loft insulation.

FITTING THE PIPEWORK

Avoid local high points in pipe runs where air could become trapped. Using plastic pipes might cause major injuries, as they will almost certainly explode under the combination of temperature and pressure generated. Where O-rings have to be used or replaced, use only solar-quality O-rings, rather than traditional plumbing O-rings, which will melt at the temperatures experienced inside the thermal loop.

The installer has to start somewhere, and so Justin began fixing the Armaflex insulated stainless steel pipe at the pumping station end.

This section was extended and clipped in position on the rafters ...

... until the point where the relevant pipe from the collectors came through the roof.

It is necessary to use suitably robust fixings because, once the pipework is full of water and anti-freeze, it will have a considerable weight.

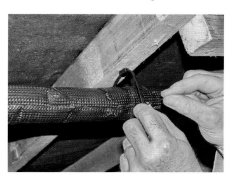

These fixings enabled Justin to hold the pipe and insulation in place with cable ties. There's no need to pull the ties so tight that they compress the insulation.

Justin re-inserted the stainless steel pipe through the hole in the roof – we saw it being fitted in Part 3 – and prepared to connect the pipes with a 90° elbow before completing the insulation.

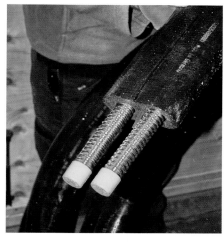

Where two pipes run side-by-side, they can be purchased as a pre-insulated pair.

Here's how you can make a union nut connection to stainless steel solar pipes using Intaeco parts, without using a flanging tool. Begin by cutting the pipe with an Intalnox (or similar) pipe cutter. Do not use a hacksaw – it won't be possible to make a smooth, perfectly square cut, and there will be filings and swarf.

Slide onto the pipe a ¾in (19mm) nut (or a union with a male thread, if required), then place the special C-clip around the pipe, immediately behind the first complete segment ...

... and carefully close it without over-tightening or distorting it.

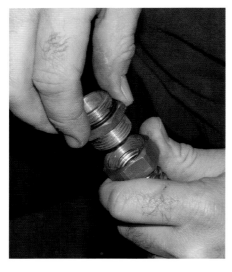

Select any ¾in (or metric equivalent) flat-face solar fitting connector WITHOUT the gasket.

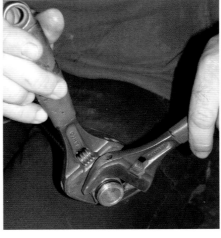

Tighten the nut that you earlier slid onto the pipe against the fitting, as tightly as possible. This action forms the flange to create a sealing 'flat.'

Undo and remove the fitting. Closely inspect the flat created on the pipe. Look for any imperfections, which may require the operation to be repeated. If the flat created is okay, place the gasket on the flat and tighten the nut against the fitting to create a sealed joint.

There is also an Intaeco flanging tool which is designed to be used in conjunction with the above fittings.

On this occasion, the Intaeco fittings were used to connect to the 3-way valve block at the pumping station.

Here is another way of creating a flange on the end of a piece of stainless steel solar pipe of this type.

After cutting the end with the correct tool, as described earlier, the pipe is inserted into the tool and clamped in place.

The plunger or ram inside the tool is forcibly slid on & off the end of the pipe several times ...

HOT WATER CYLINDER

• TIP: Note that the effective heat storage capacity of a hot water cylinder can be increased in summer weather by setting the maximum temperature higher than would usually be the case, and using a tempering valve (a mixing valve that can be set to one of a range of temperatures) to mix the hot water with cold before delivery to the hot taps, to prevent scalding.

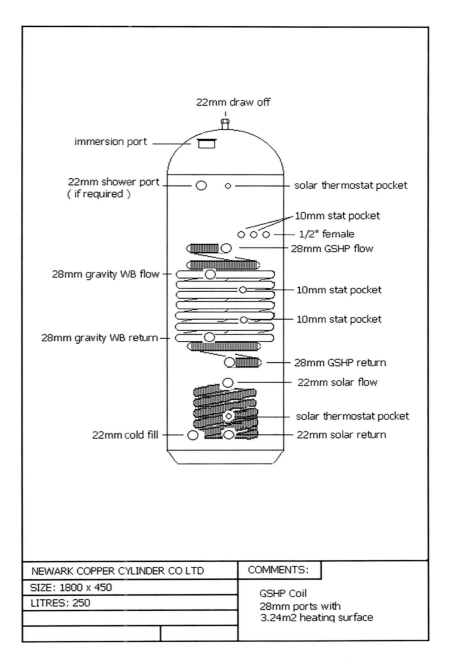

... forming a flange similar to the one described earlier. Once again, don't forget to fit the nut first or you'll have to cut off the flange and start again!

At this point, Julian has attached the 3-way valve to the green-bodied operating mechanism, and you can see how a very short connecting length of flange pipe has been fitted before the valve assembly was installed beneath the pumping station.

The rest of the pipework was then connected, including that to the heat dump (aka radiators) in the loft.

This is the specification of the tank that we installed, as received from the manufacturer, Newark Copper Cylinder Ltd. Note that this cylinder has three coils, an immersion heater port, a choice of several outlets and pockets for heat sensors.

49

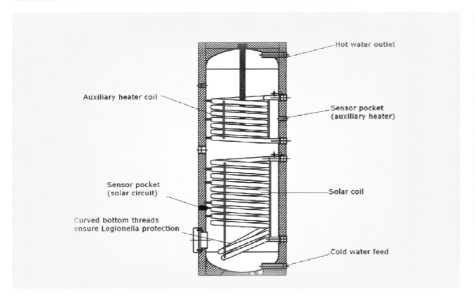

... before fitting the immersion heater in the usual way ...

This hot water cylinder differs only in that the cylinder has a convex bottom, so the solar coil is specially angled at the bottom. Depending on the installation, it may be preferable to have a supplementary water heating coil in this location if the prime objective is to eliminate any possibility of legionella (the bacterium which flourishes in air conditioning and central heating systems, and causes Legionnaires' disease) in the bottom of the tank.

HEAT SENSORS

Eco-Nomical points out that these sensors are not thermocouples (which produce a voltage, the magnitude of which depends on temperature) but thermistors, which vary in resistance according to temperature. Because of this, their connecting wires may be easily extended to at least 50m (160ft), provided the resistance of the wire used is not significant. Use wire, preferably stranded, of at least 1mm^2 cross-section, and solder the joints. Obviously, ensure that the joint is protected from damp, and that the cables are electrically insulated from each other and the environment. Do not run the wire in close proximity to power cables, as this will cause incorrect temperature readings.

On the Newark tank, the positions for the heat sensors are indicated by bumps in the insulation on the tank. The foam bump is broken off and the location of the sensor pocket has to be established with a tool such as a screwdriver.

... making absolutely certain that the sealing washer was in place ...

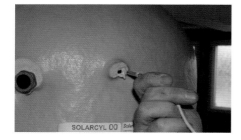

The pocket is sealed, so it is only necessary to wipe heat transfer paste onto the sensor before inserting it into its pocket.

Justin also cut away excess foam from around the immersion heater port ...

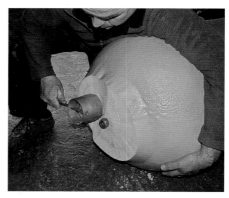

... then tightening with the correct purpose-made tool. Note the way in which the threads for the top draw-off port, as well as all other threaded ports on the cylinder, have been cleaned.

A hot water cylinder should not sit directly on the floor, so a wooden support base was constructed first, following the instructions from Newark Cylinders.

Once in position, Justin continued with connecting the solar pipework.

He next moved on to the plumbing side of things. Note the way in which each connection has been marked in advance to avoid mistakes.

Back in the loft, with the final part of the as-yet-uninsulated pipe temporarily suspended on a cable tie, Justin connected to the collectors, and also to the emergency overflow valve that is used to dump the water and anti-freeze out of the system in the event of excessive overheating, in order to prevent any risk of over-pressurising. Where the pipes penetrate the roof, insulation is not typically passed through the hole, but butted up on each side. This is a trade-off between good insulation practice and ease of sealing the roof against the elements.

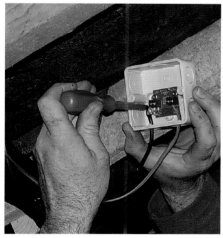

As a qualified electrician, Justin was also able to make the power connections for this valve. Note that, in the UK, electrical work has been more strictly regulated since later building regulations came into force. While it has always been a requirement to adhere to wiring regulations, now, many domestic electrical modifications require approval from a building control officer. It would appear that adding a fused spur for a pump/controller, provided it is not in a kitchen, bathroom, or other special location, does not seem to require notification to building control. However, if in any doubt, consult your local authority's building control department.

Similarly, Justin was able to connect up the controller and display panel which is situated in a bedroom or other convenient location.

SOLAR CONTROLLER – HOW IT WORKS

It is helpful to understand the operation of the solar controller, which controls the pump. A temperature sensor is placed in the sensor housing in the collector panel, and another sensor is placed in the lower level of the hot water cylinder. The controller switches on the pump if there is sufficient temperature difference between the water in the collector panel and that in the hot water cylinder. When using Resol controllers (the type used here), the installer should use the sensor with the black wire in the collector, because the lead has a high temperature silicon insulation which can cope with the temperatures that can develop there. In the event of lightning striking a collector, a voltage surge could travel down the sensor wire and be likely to damage the controller. The use of an anti-surge device, connected in series with the sensor wire, will usually prevent this, although it is important that it is connected so that the connection marked with an incoming arrow is connected to the sensor, and that with an outgoing arrow is connected to the controller.

FILLING THE CIRCUIT

The solar circuit must be filled with a mixture of water and anti-freeze. Because of its proximity to domestic hot water, non-toxic anti-freeze must be used, and the most common is a pre-mixed solution of propylene glycol, corrosion and other inhibitors, and de-ionised water. When the solar heating system is filled, the heat transfer medium (glycol) displaces the air within it. However carefully the system is filled manually, small bubbles of air will be caught in 'air locks' inside the pipes. Also, some of these air locks will be from air naturally dissolved in the water within the glycol solution. An auto-bleeding system, as described elsewhere in this section, reduces the incidence of air locks, and saves time bleeding the system.

Two valves, with a further valve between them, are fitted to the circuit; in this instance within the pumping station, to allow for a filling/draining loop.

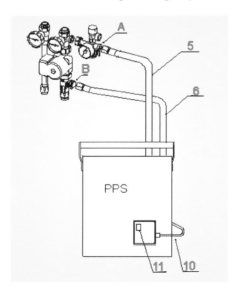

This drawing shows the filling pump connecting to the double pump station.

In order to fill the circuit with the anti-freeze mixture, some sort of pressurized filling device is required. According to Eco-Nomical, this can be made from a garden sprayer, by simply replacing the lance with a connection which fits the filling valve. The sprayer is connected to the (upper) filling valve, the adjacent drain valve is opened, and the valve between the two is closed. The anti-freeze mixture goes in the sprayer reservoir, the sprayer is pressurized, and the filling valve is opened. When the anti-freeze mixture starts to come out of the drain valve, close the drain valve and increase the pressure to around 2.5 bar. Then open the intermediate valve. It may be necessary to do this in more than one step if the capacity of the solar loop is greater than that of the sprayer reservoir. When the pump is running, slacken the large screw in the centre of the face of the pump to bleed the pump, then re-tighten. Ensuring that there is sufficient anti-freeze solution in the sprayer reservoir, repressurise to about 1.5-2.5 bar, and open the automatic air bleed valve at the top of the circuit. (See page 53.) Typically, this valve is left open for a few days until all the air has bled from the system.

Place the filling pump station in a secure, horizontally clear place, so the hoses can easily connect to it. Fill the container with glycol mix.

On this system (see left pic), the flow hose (5) is connected on the part of the double pump station where the safety relief valve (A) is located. The return hose (6) connects to the valve below the Grundfos pump next to the flow setter (B). Switch on the filling pump station and open valves A and B, or, in the case of the installation being carried out here, follow the instructions for operating the valves.

Be prepared to carefully add additional glycol, as required, to the tray while it is operating. If the installer lets it run out, air will be reintroduced to the system.

The air vent should have been fitted at the highest point in the circuit to allow it to be purged of air. Justin opened this valve so that, while pumping was taking place, air was purged from the system. He recommends closing the valve once again after the first few weeks of running, and allowing the de-aerator (see page 44, top-right) to take care of the small amount of day-to-day aeration that can occur.

It's easy to forget, but Justin remembered to bleed the heat dump radiator valves.

Flow settings

These must be set in accordance with manufacturer's specifications. For instance, the ATAG flow regulator should be set to 1.0 litre per minute per panel. Its three-panel installation is set to a flow rate of 3.0 litres per minute.

Pressure

Once the system has been pressurised, ATAG's advice is to keep it between 2.0 and 3.0 bar. With this pressure, the glycol/mix will not evaporate until a temperature of around 120°C is reached.

PART 5: MAINTENANCE

There are many safety controls built into every system, and ATAG, as other manufacturers, relies on a number of safety strategies. When these are followed, a properly-engineered thermal solar system is as safe (if not safer) for the consumer than when using a pressurised central heating system in the home to heat water.

Maintenance requirements for solar installations

Note: An appropriate written record must be kept. ATAG recommends that every installation should have a specific anti-legionella regime matching the installing heating and ventilation engineer's and/or the system designer's proposal.

Maintenance checks

The following checks are designed by ATAG for large-scale installations, but are a useful, comprehensive guide for any solar thermal system.

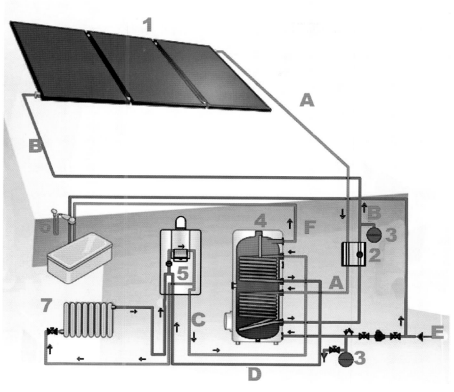

1) 1000-10 ATAG solar panel, 2) Double pump station, 3) Expansion vessel, 4) Hot water cylinder, 5) Auxiliary heater, 6) Hot water domestic supply, 7) Heating system
A) Flow pipe (hot Glycol) on solar system, B) Return pipe (cold Glycol) on solar system, C) Flow pipe from auxiliary heater, D) Return pipe to auxiliary heater, E) Cold water feed, F) Hot water supply to tap

October to March: every sixty days.
April to September: every ninety days.
- Visual: weeps, leaks and fluid escapes.
- Manually operate all valves and pumps to ensure that they are working correctly.
- Visual: system pressure has not dropped into red zone.
- Visual: flow gauges indicate correct flow levels.
- Visual from ground level (with binoculars), for damage to panels or fixings.

Annual checks –
- Operation of active or non-active anodes.

Every three years –
- Pre-charge of the expansion vessel(s).

- Whole system for signs of corrosion.
- Test glycol with approved testing kit.
- Visual: all insulation is in good condition and not becoming brittle or cracked.
- Visual: pre-heat cylinders are free from corrosion.
- If vacuum panels are used, check that the vacuum gauge shows there is still sufficient vacuum. The pointer should be in the green zone.

If case of any unexpected changes, contact the installer immediately. Note that fluctuations in pressure readings while the system is operating are perfectly normal.

Chapter 5
Wood biomass central heating boilers

INTRODUCTION

Biomass fuel is made from organic matter, which, in principle, could be anything from bales of straw to animal carcasses. In practice, though – at least as far as this manual is concerned – it relates only to wood biomass. Like all biomass fuel, it is essentially carbon-neutral: if the fuel was left to decay or disposed of in any other way, it would put the same amount of carbon into the atmosphere as when it is burned. Moreover, when the fuel is replaced, the new growth takes out of the atmosphere the same amount that has been 'used' when the fuel was burned.

Of course, in reality, life is never that simple! Some carbon dioxide will be emitted in the production of biomass and in its transportation. What's more, if old woodland is used for biomass fuel, it can take decades or even centuries to replace the carbon

through wood growth. And if woodland is replaced by other plants with lower carbon uptake, there is a net increase in carbon dioxide in the environment. These are environmental issues that should not be overshadowed by wishful thinking that, because biomass fuel *can* be carbon neutral, it *will* necessarily be carbon neutral. However, it does have that potential in a way that the burning of fossil fuels does not. A pdf publication available online (at the time of writing) from Friends of the *Earth entitled Energy from biomass: Straw Man or Future Fuel*? neatly sums up the argument.

It could be said that biomass fuel is the oldest type of heating known to man, but modern technology is such that the efficiency of creating domestic heat from biomass is somewhat better than when flints were struck to start the first fires. It's a technology that requires specialist knowledge, and

this section of the manual has been produced with the assistance of, and information from, Euroheat, which has become the UK's leading wood burning stove and wood biomass boiler company. Euroheat has provided some wonderful illustrations for use in the manual, and one of the reasons that these are so effective is because the company has a strong appreciation of the benefits of education. Euroheat has created the country's first wood biomass training centre, accredited by the government-recognised Heating Equipment Testing and Approval Scheme (Hetas), at which members of the heating industry and the public can see wood biomass boilers in operation, and learn how they work. In fact, almost all of the following information is derived from information supplied by Euroheat – a testament to the extent of that company's expertise.

PART 1: WOOD FUELLED APPLIANCES – OVERVIEW

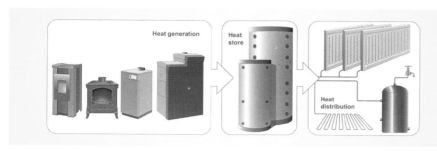

When burning wood, there are two stages in the combustion process: firing the volatile gas and burning the fixed carbon (charcoal). Wood burning boilers, in common with all wood biomass fuels, burn most efficiently and cleanly when burning hot and fast. Many years of development work have made modern wood burning equipment, such as those available from Euroheat, extremely efficient, enabling them to utilise nearly all of the energy in the wood.

- The release of gas, known as gasification, takes place when wood part-burns in the fuelling chamber, releasing volatile gas.
- This gas is drawn into a separate combustion chamber where it mixes automatically with secondary air to burn in ideal conditions.
- Typically, the on-board system control monitors the oxygen concentration and/or temperature of the exhaust gases or water temperature.
- The controller automatically adjusts primary and secondary combustion air through independent air controls, optimising combustion as the wood burns. (Euroheat)

Pellet and wood chip appliances

Both pellet and wood chip appliances can offer a level of control similar to that of oil and gas appliances: the heat generated can be quickly switched on and off. This is one of Euroheat's pellet-powered boilers. (Euroheat)

With both pellet and wood types –
- Ignition is achieved automatically.
- A transfer system carries the fuel to the combustion chamber.
- A sensor regulates a measured air supply, and the heating output adjusts itself continuously according to need.
- The appliance will clean itself and

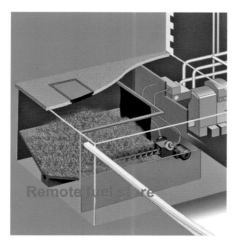

Wood chip boilers are generally more suitable for larger domestic properties and commercial applications because of the requirement for a large, dedicated fuel store. (Euroheat)

shut down when energy is not required.

Log burning stoves for space heating
The simplest example of burning wood biomass fuel is burning logs in a stove acting as a room heater only. If a chimney is not available it

can often be constructed without too much disturbance. Dedicated log-fuelled boilers can sometimes provide full-house heating, even in large properties.

Wood burning stoves are generally used as a secondary heat source, but in open-plan areas can heat a large proportion of a property.

These stoves operate in three different ways –

Intermittent operation – a stove which is designed for quick heating for short periods, such as for evening use.

Heat storage – a stove which is often constructed with a stone or ceramic exterior or internal stone mass to collect heat quickly from intermittent operation, and then release the stored energy slowly over a period of time.

Continuous operation – an advanced appliance designed with very accurate combustion air control over the fire's activity. This control allows for long periods of operation without user input, such as overnight burning.

The pros and cons –
- Aesthetically satisfying – no form of heating is more comforting than the sight, smell and feel of a log fire.
- Some models approved for use in 'smokeless' areas.
- Operates without a demand for electricity.
- An (usually optional) external combustion air connection improves property insulation.

- Logs are bulky in nature, require correct storage methods, and need to be transported by hand.
- Logs need to be correctly dried before burning.
- Stoves need to be tended to and replenished regularly when in use.
- Regular cleaning and ash removal necessary.

Log burning stoves for domestic hot water and central heating
A stove with a boiler can be used to run or supplement your central heating and/or hot water. It can operate with a closed 'sealed' or gravity 'open vented' circuit, and in combination with alternative heating solutions. A water heating stove is the ideal supplement for resource-saving while at the same time serving as a warming and attractive room heater. Because heat output is divided between the water

and the room, it is very important to choose the right model to achieve the correct balance, which will be influenced by anticipated demand of the heating system as well as room size.

The pros and cons –
- Centralized home heating from one location.
- Advanced modern stoves have an improved ratio of water to room heating (up to 70/30%).
- Can be linked to alternative heating sources to provide water heating.

- Much larger quantities of dry wood will be required for central heating than for a room-only heater, though the difference will not be that great if the boiler is heating domestic hot water only.
- More cleaning will be required. The stove creates ash which has to be removed, leading to dust, and regular carrying of wood into the house creates dirt.
- Quantities of dry, ready-to-use wood require storage close to the property.
- Suitably-sized log stores will be required to dry wood during the summer.

Log burning boilers

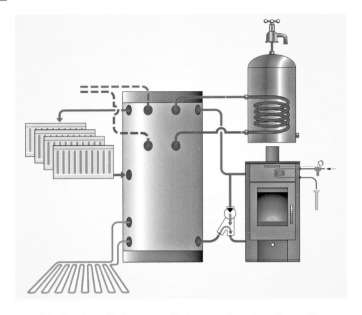

Modern installations usually have a closed system with expansion vessel and thermal store. Closed systems are un-vented, pressurised, closed to the atmosphere, and therefore have no feed or expansion tank. It is important to include the appropriate safety control features to deal with any water expansion or excess heat generated. Only suitable for modern stoves fitted with a thermal safety device. Shown here with thermal store, which enables heated water to be retained for use when its energy is needed later. (Euroheat)

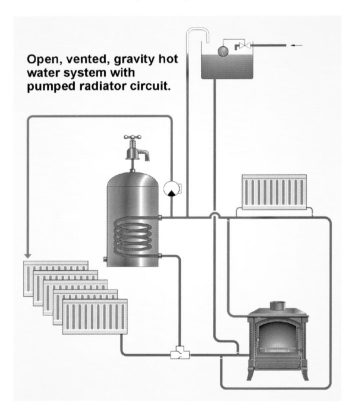

Gravity feed systems are vented to atmosphere and require an expansion tank, usually situated in the loft space. A way of dissipating excess heat in case of an electricity or system failure is required. This is normally an adequate 'heat leak' (or 'heat dump') in the form of a permanently 'on' radiator, often located in a bathroom. (Euroheat)

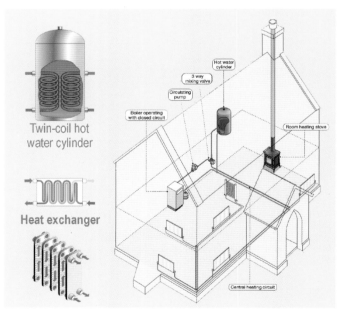

This is a closed system providing hot water and central heating, supplemented by a room heating stove. Dual coil hot water cylinders allow for two independent heating sources from different systems to perform a common service. Plate heat exchangers allow two water circuits to transfer heat between themselves while remaining separate. (Euroheat)

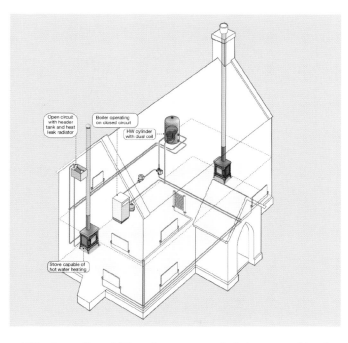

This shows the addition of an open vented stove, capable of contributing to a house's hot water requirements by utilising a twin coil cylinder. (Euroheat)

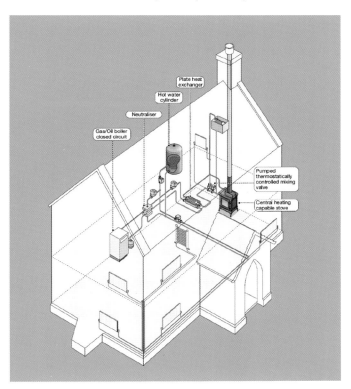

This is a stove capable of running central heating, combined with a fossil fuel boiler, with a heat exchanger to transfer heat between open and closed heating circuits. Minimum stove temperature is maintained by a thermostatically-controlled mixing valve. (Euroheat)

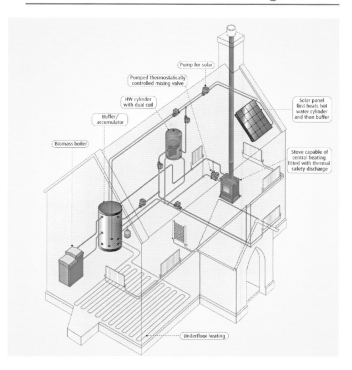

Biomass boiler and stove, which is suitable for use in a closed heating system linked via a thermal store/accumulator. Solar heating is set up to first heat the hot water cylinder then accumulator. (Euroheat)

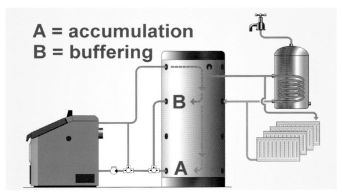

Use of a thermal store as a system linking solution from which hot water and central heating will be distributed. During the combustion process, more heated water will be produced than is required at that time. This excess heated water is transferred and stored in an accumulator – a large storage vessel. This accumulated energy can be stored for long periods until needed for heating or hot water requirements. (Euroheat)

Log boilers are fuelled by hand. When sized correctly, the fuel chamber should be large enough to hold enough wood to supply average daily heating requirements in winter. Occasionally during severe weather a second load may be required. When choosing a log boiler, the fuelling chamber size is the main consideration, before the kW output of the boiler.

On the face of it, it is difficult to regulate the heat

output of these boilers since they are only efficient and clean when burning hot and fast, and are less efficient on tick over. Consequently, when the boiler is running, more energy (heated water) will be produced than is required, but the solution is to transfer excess heated water into an accumulator. This large vessel contains accumulated energy which can be stored for long periods until it is needed for heating or hot water.

Pros and cons –
- Simple to operate and ignite.
- Burns logs and clean wood waste.
- If correctly sized, needs fuelling only once or twice per day.
- Very high efficiency (up to 92%, depending on model).
- Can be easily linked with an alternative heating source.

- Requires a dedicated boiler room capable of housing the boiler and accumulator.
- Requires daily fuelling, by hand.
- Requires ash removal.

ABOUT LOGS

Hardwoods, sometimes referred to as broad leaf, are slow-growing and deciduous. Softwoods, sometimes referred to as conifers, are fast-growing and evergreen. Both hard- and softwoods have broadly similar calorific value per kg, but the density of softwood can be about half that of hardwood, which means that twice the amount of softwood is required in volume terms to produce the same heat output as hardwood.

Seasoning

Newly-harvested wood contains a naturally high amount of water, which must be greatly reduced, or else the heat produced must first remove the water before it can generate more heat. Logs should always have a moisture content of 20% wet scale or less, before burning.

Storage

For best results logs should be split to allow the moisture to escape more easily, and stored off the ground in a dry, covered space with plenty of air circulation. As a guide a 6in (150mm) unsplit log will take at least two years to dry – but split the wood and it is likely it will dry out in a year.

AVERAGE HEAT AVAILABLE (IN KILOWATT-HOURS) FROM DIFFERENT TIMBER TYPES

Bear in mind that the following typical values are based on the weight of each wood type. The volume – the amount of storage space each will need – will be very different. One kg of spruce or pine takes up a lot more space than 1kg of oak, for instance – and the lighter woods burn away much more rapidly, creating more ash.

Wood	kWh/kg
Spruce	4.67
Beech	4.13
Pine	4.50
Fir	4.62
Oak	4.33
Ash	4.21

PELLET STOVE ROOM HEATERS

Pros and cons –
- High degree of control.
- Clear view of the fire.
- Designed to be highly efficient.
- Easy to fuel with lengthy period between refuelling and cleaning, leading to less dust and dirt.
- A natural fuel, widely available.
- Only small storage area required for pellets.

- Good pellet quality is important.
- Requires an electrical supply.
- Flame appearance not as attractive as that of a log burning stove.
- Some background noise may be detectable from the operating mechanism.

WOOD PELLET BOILERS

These boilers (see page 53) function almost like oil/gas-fuelled boilers. When required they automatically ignite, feed fuel to the fire, self-clean, and switch off when not required.

Almost any heating system can be retrofitted with a wood pellet boiler, either as a stand-alone heat source, or as a primary or back-up heat source in conjunction with another boiler.

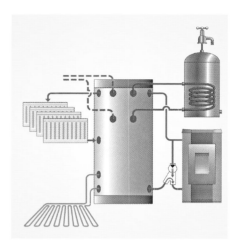

Pellet stoves can be left to burn all day with minimal attendance. Heat output is monitored by the in-built controller regulating the flow of pellets into the combustion chamber via a screw-feed mechanism. The flame is not the same as with a log burning stove: it's much more consistent, though a little less aesthetically pleasing. (Euroheat)

Wood pellet stoves offer similar aesthetics to a wood burning stove without most of the disadvantages of logs. You only need to push a button or set the timer and the controller automatically regulates the burning operation. Wood pellet stoves burn very efficiently and, because pellet fuel is more energy-dense than logs, the pellets require far less manual handling. The hopper of the stove can sometimes hold enough pellets for several days' burning, depending on model. (Euroheat)

ABOUT PELLETS

Wood pellets are a CO_2-neutral fuel (apart from transportation costs, and some are sourced in different continents), typically made from compressed sawdust, and should be from renewable sources. They are small, dry, uniform, energy-dense and 'flow' easily. A kilogramme of pellets has roughly the same energy content as half a litre of heating oil.

Pellets are available in pre-packed bags (normally 10 or 15kg) and can be bulk delivered in separate bags, on palletts or 'loose.'

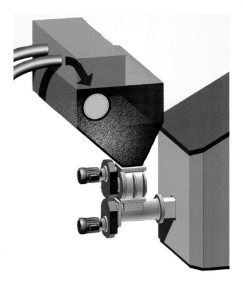

Larger domestic models can be vacuum-fed from an adjacent separate store. For bulk deliveries, pellets can be blown directly into a large, on-site fuel storage area, then transferred to the boiler automatically, either by vacuum ...

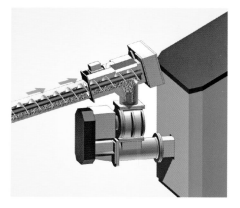

... or via an auger-feed. (Euroheat)

Storage

The provision of a dry store is required as it is important to keep the pellets from becoming wet during storage. Where an automatic feed is installed, a bunker is required from where the boiler feed system can draw its supply, and this should be positioned where deliveries can easily be made. The size of the store should be calculated to allow for the demand of the boiler and the frequency of delivery. 600-650kg pellets = $1m^3$.

Pellet quality

A good quality pellet is very important. It will have a surface that appears smooth and shiny, uniform length and no dust. Poor quality pellets have longitudinal cracks, a high proportion of extremely long or short pellets and a high dust content.

Pros and cons –
- Fully automatic control/feed systems.
- Automatic ignition and cleaning.
- Less bulky fuel, and cleaner than log or wood chips.
- Boilers suitable for all size applications.
- Designed for long operational life.

- Pellet quality is important for good operation.
- Higher installation cost compared with fossil fuel boilers.
- Requires dedicated boiler house and fuel store.
- Requires electrical supply.
- Can require more maintenance than fossil boiler.
- Requires occasional ash removal and inspection.

Wood pellet safety: explosion risk

Wood pellets can create dust, which poses a safety risk. Pellets can break up and revert to sawdust when blown into a silo and it is possible for an explosive mixture of dust and air to be created, so minimising dust build-up is essential. Good quality pellets have low dust content, and tend to be structurally stronger than low-quality alternatives. Similarly, the quality of the delivery system is important. Badly-designed pipework can cause pellets to erode or shatter on impact. For this reason, there must be an in-built mechanism for dust collection, and delivery drivers must avoid using excessive pressure when blowing pellets into the store (which should be regularly checked and cleaned out). Delivery pipes must be bonded to earth to avoid any build-up of static electrical charge. To prevent ignition from any electrical source, any essential electrical fittings within the store must meet the appropriate Explosives Atmospheres (ATEX) specification or local equivalent.

WOOD CHIP BOILERS

Wood chip boilers are in many ways similar to pellet boilers, except, of course, for the fuel itself: wood chips are much bulkier than pellets. Modern wood chip boilers are highly efficient, clean burning, and totally automatic. They are generally more suitable for larger domestic and commercial applications. Some boilers designed for wood chips can also burn pellets; however, boilers designed specifically for pellets cannot generally use wood chips. Euroheat's HDG chip boilers do not have to run constantly. Due to their advanced auto ignition, the boiler can run for periods as short as one hour and then switch off until next needed.

Wood chips

These are normally 2cm to 5cm in length, and are made from waste wood, brush, saplings, tree waste and standing timber from logging operations, and from forestry and roadside maintenance. Fuel growing methods, such as brush and coppice farming, can produce ideal wood for chipping on a sustainable basis with a very high yield per acre, and a short cycle. Coppicing can be mixed with conventional forestry timber.

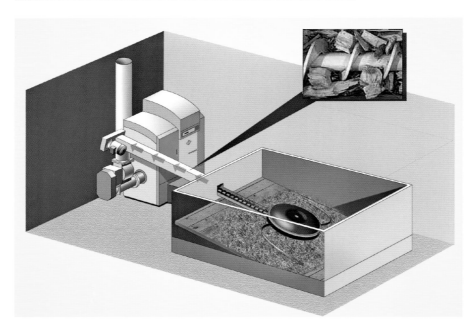

Like pellet boilers, all wood chip boiler systems share the same basic features, which comprise a boiler, a storage facility, and a feed mechanism. Wood chips are made from whole trees, branch wood or coppice products which have been mechanically chipped. Ideally, the wood should have been air-dried before chipping, or chipped then allowed to dry. Wood chips are delivered into a local bunker close to the boiler. On demand the wood chips are augured into the boiler, which maintains a constant fuel supply. Wood chip boilers must have built-in safety features such as a rotary sluice prior to the final combustion feed auger, to prevent any back burn. (Euroheat)

Delivery

Chips can be transported and unloaded by tipper truck, auger or blown delivery. Because they are generally available locally, long-distance haulage, packaging and energy consumption are all reduced.

Storage

Wood chips are typically stored in a timber-floored bunker, and should be kept under cover to prevent wetting. Good airflow is essential to disperse water vapour, minimize the chance of composting, and reduce mould formation. An auger feeds the material to the boiler. Delivery in bulk is usual for wood chips, and there has to be provision for tipping – or otherwise shifting them – into the bunker. Key considerations are access for vehicles, proximity to boiler, and frequency and method of fuel delivery.

Wood chipping

If local wood stock is available this can be chipped on site by locally available machinery or a local contractor. Wood chippers cut across the grain

via the action of a set of sharp blades set round the surface of a disc. Disc chippers have hydraulically-powered feeding rollers which draw the wood in between the cutting blades. Often the chip size is graded so that it conforms to size requirements.

Wood pellet and chip safety: toxic/harmful gas risk

The UK's Health and Safety Executive (HSE) advises the following (these are the parts relevant to domestic installations) –

- Wood pellet hoppers/tanks/storage rooms and boilers should always be installed and commissioned by a competent person, usually approved by the manufacturer/supplier.
- Do not enter the pellet storage area or place your head into a wood pellet hopper as it can contain toxic gases. No-one should enter the hopper/tank unless fully trained and competent in confined space entry procedures, as per the HSE's Code of Practice for Working in Confined Spaces. This should include adequately ventilating* the

storage area, and checking carbon monoxide and oxygen levels with an appropriate device prior to entry. It is recommended that the storeroom is ventilated at all times, either mechanically or by being designed to have a through-draught.
- Manufacturers, suppliers and distributors of wood pellets should provide adequate health and safety information to the user in their materials safety data sheet.
- Warning signs should be placed on the pellet storage area access door; ideally on both sides so that it can be seen when the door is open. The warning sign should indicate the following –
DANGER – RISK OF CARBON MONOXIDE POISONING
- There is a danger to life from odourless carbon monoxide and lack of oxygen. Check atmosphere before entry with an appropriate device. No entry for unauthorised persons. Keep children and animals away from the storeroom.
- No smoking, fires or naked flames.
- The room should be adequately ventilated before entering. Keep the door open whilst inside.
- There is a danger of injury from movable parts.

*This must be discussed with your specialist supplier because too much damp intake will destroy the integrity of wood pellets.

Pros and cons (much the same as for pellet boilers except) –
- Can make good use of local wood by using a chipping machine.

- Requires larger fuel bunker than wood pellets.
- Due to initial capital costs more suitable to larger domestic and commercial installs.
- Requires correctly-sized, properly-dried, graded fuel.
- Requires more local maintenance and inspection.

When wood pellets are stored, they can create carbon monoxide (CO) gas. Carbon monoxide can kill quickly and without warning. It is a colourless, odourless and tasteless gas that is highly toxic. When carbon monoxide enters the body, it prevents the blood from bringing oxygen to cells, tissues and organs. For that reason, nobody should enter an enclosed pellet store alone (there must always be another responsible person in attendance outside the store), and the store should not be entered until it has been properly ventilated. Similarly, dust must be carefully contained within the store and not allowed to escape into adjacent areas, in particular the boiler room which, by definition, will always have a point of ignition. There may need to be a fireproof partition between the fuel store and the boiler room: to be decided upon by your specialist supplier.

COMBINED SYSTEMS

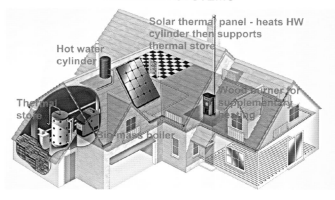

Solar thermal panel - heats HW cylinder then supports thermal store

Hot water cylinder

Thermal store

Wood burner for supplementary heating

Biomass boiler

It's not unusual for users of biomass heating to also install one or more other form of renewable energy. Thermal solar domestic hot water heating can also be integrated into the biomass home heating system. (Euroheat)

PART 2: INSTALLING A BIOMASS BOILER

Although the details will vary for the installation of different types of biomass boiler, and in the different types of location in which they can be situated, this procedure, followed by the Efficient Energy Centre, as shown here installing a SHT Thermodual TDA 40 in the home of Mr Bullough, provides a good basis for understanding the installation of any similar biomass boiler system.

Euroheat's SHT TDA Thermodual biomass boiler has the advantage that it can run on either logs or pellets. First introduced in 2004, the boiler comes in 15, 25, 30 and 40kW formats, and takes split logs of up to about a third of a metre in length. This boiler allows the use of firewood while eliminating the need for constant attention and refuelling to maintain heating capacity.

In the installation shown here, Efficient Energy Centre had several fully qualified installers working on the same installation at once. In order to make the installation more comprehensible, I have grouped together sections of work rather than showing the strict order in which the photographs were taken. For that reason, you will sometimes see things happening in the background that appear to be out of sequence.

PART 3 INSTALLING A WOOD BURNING ROOM HEATER AND CHIMNEY

It's undoubtedly true that pure solar energy is among the most carbon-neutral of all fuels. And it's not too fanciful

Continued page 68

Mr Bullough's home is a large, former vicarage, which had been costing a fortune to heat with central heating oil while producing a relatively large amount of CO_2.

Mr Bullough has access to a ready supply of logs, which made the SHT TDA Thermodual biomass boiler particularly suitable for his installation. The Thermodual is designed to always burn the cheapest fuel available: it fires up on pellets, switches to logs, then once the logs have been consumed – usually after about five to seven hours – it automatically switches from firewood to wood pellets again, but only if it hasn't been refuelled with logs.

As is often the case with large, old houses, the former vicarage has outbuildings and barns, one of which was timber-lined and given a concrete floor to make it suitable for the biomass boiler installation. An identical amount of space in the next bay was also prepared and allocated for fuel storage. (There will always be a requirement for fuel storage when installing biomass, as discussed earlier.) Incidentally, it is important that log fuel has less than 20% moisture content when burned; it must be stored properly in order for it to dry out.

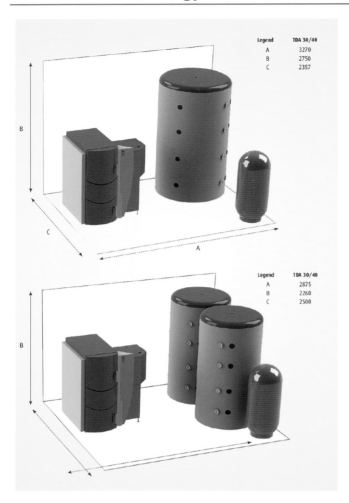

Legend	TDA 30/40
A	3270
B	2750
C	2357

Legend	TDA 30/40
A	2875
B	2260
C	2500

Wood pellet, wood chip and wood log biomass boilers, plus their ancillaries, take up considerably more room than most other forms of alternative energy. These diagrams from SHT TDA indicate the amount of room necessary for one of these two alternative layouts. The components themselves will be described later. (Euroheat)

In very many cases, biomass boilers are fitted away from the house, and it then becomes necessary to run insulated pipes underground to carry the hot water flow and return.

In order to install a biomass boiler, there must be room for the boiler and fuel, as well as sufficient access so that it is physically possible to deliver the boiler in the first place, and so that fuel can be delivered in bulk quantities, if required.

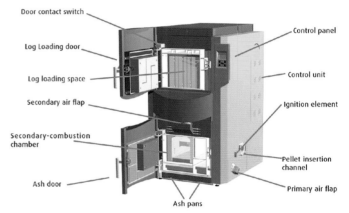

These are the components of the SHT TDA Thermodual boiler.

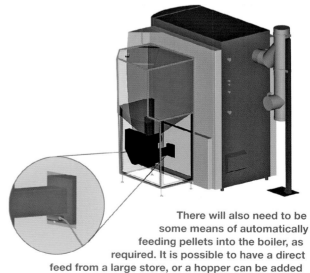

There will also need to be some means of automatically feeding pellets into the boiler, as required. It is possible to have a direct feed from a large store, or a hopper can be added to the side of the unit (the area outlined in red, on the left of the boiler), which means that the pellets have to be manually loaded into the hopper every few days. (Euroheat)

There must also be room for the 750kg (1650lb) boiler to be off-loaded from the delivery vehicle into a position that allows access to its final location.

The guys from Efficient Energy Centre oversaw the off-loading …

… before using a pallet truck to manoeuvre the boiler, complete with its ready-fitted hopper seen here on the right of the boiler, into position.

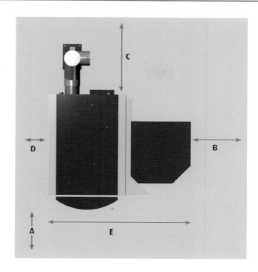

Boilers, of course, vary in size, and different manufacturers will have different recommendations, but these minimum clearance dimensions, provided by SHT TDA, give a good indication of what is usually required for other systems, too. (Euroheat)

Measurement (mm)	Description	SHT TDA 30/40
A	Wall to front	650
B	Wall to right	300
C	Wall to rear	500
D	Wall to left	100
E	Width with manual-fill pellet hopper	1420
E	Width with intermediate pellet hopper	1325
	Minimum room width with manual-fill pellet hopper	1820
	intermediate pellet hopper	1725
	Minimum room depth	2357
	Minimum room height	2100

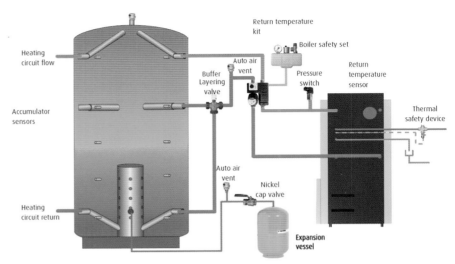

Of course, the foregoing dimensions do not include the accumulator or expansion vessel (which is being carried into position here). If siting the accumulator in the boiler room, clearance dimensions will need to be increased respectively (including room height). For this particular installation, see page 61, 4th picture.

This schematic drawing from SHT gives an indication of the type of pipework required. (Euroheat)

While the main components were being placed in their approximate positions, Owen Pugh began laying out the fixtures and fittings in orderly groups so that they could be found and fitted instantly while installation was taking place.

One of the first jobs was to locate the flow and return pipes where they entered the new floor. The installation sleeving can be clearly seen here. You can also see that the ends had been capped before the floor was laid which, in view of the inevitable amount of concrete splattering that takes place, is an essential step.

Ed Udall used compression fittings to attach taps …

… to both pipes before extending the pipe runs vertically …

… and attaching them to the wall using a commonly available type of industrial fixing known as a Unistrut slotted steel support channel.

Later on, the pipes were fully installed, including the return seen here, connected into the bottom of the accumulator.

This installation was simpler in some ways because of the timber lining on the walls, which made it easier to connect the support channel and their ancillary fixings. Usually, there's a lot of drilling of masonry and use of wall plugs. This is the circulation pump installed in the flow circuit …

... while above it another tap was fitted so that the pump could easily be isolated, should it require maintenance in future.

Steve then began to install the return temperature kit. He had to ensure that the pump was mounted vertically to prevent any pockets of air from being trapped inside.

... and, when the power is operated, the jaws clamp the fitting to make a foolproof, water-tight joint in seconds.

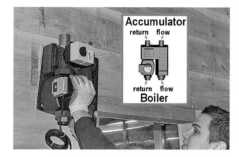

Steve Crowe held up the return temperature kit (which fits between the boiler and accumulator) in order to establish where he wanted it to be located. It had to be in a place that would be appropriate for the pipe runs to be fitted later, as shown in the diagram (inset).

Now, the installation of the pipework for that section of the system could commence. Heavy-duty pipe cutters are required to cut through the steel pipe.

The pipe had been cut over-length, so Steve cut it to length now, in situ ...

... before the first elbow was added. You can see here the appearance of the inside of the elbow.

He marked out, using a spirit level to check accuracy ...

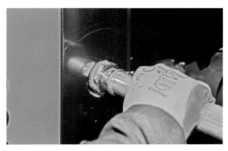

The Efficient Energy Centre's engineers use a very expensive but quick and reliable system of pipe joining. Here, a special fitting has been screwed onto the back of the boiler, and the end of the pipe pushed fully home.

... then screwed the backing plate into position on the wall.

A Geberit Mapress press-fitting tool has been placed on the fitting ...

After being pushed fully home ...

… the next joint in the sequence is made as you can see here, where Steve has started to work on the return circuit to the boiler.

There must be no shut-off valves between the boiler safety set and the boiler. Note again the use of a spirit level to ensure correct alignment.

The correct specification is to use soldered connections where the cold water feed connects to the thermal safety device on the side of the boiler.

It's now obvious why major components are installed first to provide fixed points …

This safety gear is used to operate an integrated water extinguishing system to provide total safety in the event of a catastrophic failure elsewhere. This one is encased in its own insulated box.

The flue starter kit was fitted onto the flue spigot on the rear of the boiler. It is most important to ensure an airtight seal using suitable high temperature sealant. As with any biomass system, it is essential that the flue is efficient and correctly installed, to ensure that (potentially dangerous) gases cannot escape from the flue …

… to and from which the new pipework can be connected.

While Chris Tyler had been installing the safety gear, Steve Crowe continued to run the pipework from boiler to accumulator.

… and because well-insulated pipes and appropriate pipe runs, with a minimal number of elbows, are essential for the efficient operation of the boiler.

The next step is to install the boiler safety set on a branch on the return circuit.

Even though the circuit is pumped, it's still good practice to ensure that pipework is installed so that there is no possibility of an airlock forming within it.

There are some similarities between fitting a new flue through the inside of a building and installing water pipework, in that the installer has to get from 'here' to 'there' in a way that doesn't clash with the rest of the installation, but also while preserving a safe distance between the flue and any flammable building materials.

The lower end of the flue starter kit was assembled on the ground ...

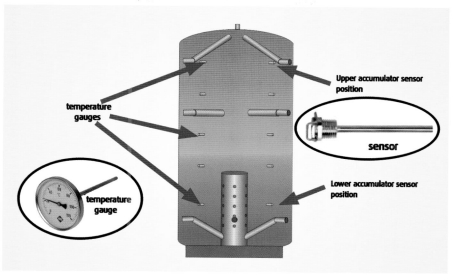

Back down to earth, and the accumulator has to be fitted with a temperature gauge and sensors, installed in pockets built into the sides of the accumulator. (Euroheat)

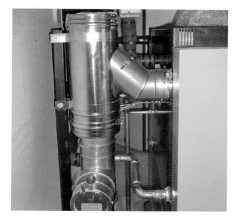

... before being added to the flue on the back of the boiler. Note the support stand on the far (left) which is part of the flue starter kit. The flue starter kit acts as a condensate tap and has a draught stabiliser built into it which can be set to give the correct draught for combustion through the chimney.

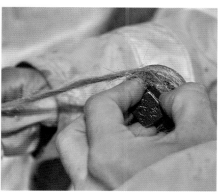

Where black iron fittings are screwed into threads in the accumulator, so-called horsehair (actually hemp), is wound in a clockwise direction into the threads with sealing paste ...

And experience also tells him how tight is tight enough without the risk (especially on copper cylinders) of tearing the built-in thread out of the cylinder through attempting to grossly over-tighten the fitting.

Strict UK building regulations apply to the passage of a flue pipe near to potentially flammable materials such as roof joists, and it's also essential that the height and location of the chimney are correctly calculated so as to avoid any risk of downdraught.

... so that when the fitting is screwed in place, the seal is both tight and water-tight. Experience tells the fitter how much hemp to use.

At the end of the first day, the Efficient Energy Centre guys had done a great job of getting the major components and some of the connections in place.

The next day, pipework was connected to the expansion vessel, and once all the pipes were fully installed …

… the system was pressure-tested and only then, the insulation fitted in place.

Fitting insulation is an essential part of the process, bearing in mind that in every other respect, the system is intended to work at maximum efficiency. So the idea of throwing heat away through uninsulated pipework just doesn't bear thinking about!

Heating systems without inhibitors will corrode and are prone to limescale build-up and corrosion. The final job is to add a suitable inhibitor to the system.

The completed installation looks impressive and provides a good visual indication of the amount of room required for a medium-large biofuel pellet burning heating system – or at least the active part of it, not including the fuel store.

to describe heat from a log burner as solar energy for fireplaces – and here's why.

When a tree is growing, it does so largely by utilising energy from the sun. Thus, wood for burning comes from the sun's energy. However, it also involves the absorption of an amount of carbon dioxide that almost exactly matches the carbon dioxide given out when the timber is burned for firewood. (Indeed, whether wood is burned as a fuel or left to decay in the forest, it will release the same amount of carbon dioxide into the atmosphere.)

Provided that the wood used is from a sustainable source (in other words, where new saplings are planted to replace wood that is cut down), this energy source is almost completely carbon-neutral, apart from the unavoidable costs of felling, preparation, and transportation. And if talking about using a wood burning stove with logs, this is best of all because logs are almost always obtained from local suppliers.

FLUE LORE

- An efficient flue is one that extracts smoke and gases smoothly and rapidly. Unless a flue is operating efficiently, gases and smoke in the chimney lose velocity, linger in the

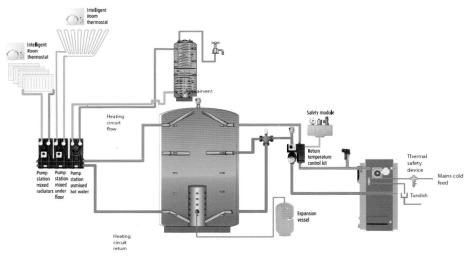

Of course, in order to be able to function, the boiler and associated components need to be connected to a conventional central heating system as shown here – but that's another story! (Euroheat)

It's important to realise that the successful installation of a wood burning stove depends as much on choosing the right stove (by which I mean one that works efficiently) as it does installing an efficient flue. Helen Davies and Jonathan Greenall, founders and owners of Clearview Stoves, provide interesting and essential advice on flue construction.

chimney space, prevent a proper draw at the fire, and increase the risk of smoke blow-back.

- A flue must be properly insulated otherwise flue gases will cool too rapidly. They need to retain their heat so that they rise up the chimney as rapidly as possible.
- The ideal flue will be vertical. Any horizontal or angled runs will slow the flow of gases.
- The inside of the flue must be as smooth as possible with no obstructions or protrusions.
- The flue's outlet must be above or away from building overhangs, trees or anything else that could cause a backdraught down the chimney.
- If there's a risk of birds nesting in or on the chimney, protect the opening with a wire cage or something similar.

IMPORTANT NOTE: There are specific regulations and important safety issues connected with flues, especially relating to the installation of a flue in connection with gas- or oil-fired devices. It is important that your builder is familiar with these matters before carrying out any work on a chimney.

LEGISLATION

The installation of a wood burning or multi-fuel stove, or any solid fuel appliance, is, at the time of writing (in the UK), controlled under Building Regulations, as is the relining or installation of flues and chimneys associated with such appliances. Therefore, it is necessary to submit an application to Building Control before starting work, unless you are employing a HETAS-approved engineer. Either route can offer the required legal certification, but note – you don't need a HETAS fitter's certificate AND approval from Building Control.

STOVE SELECTION

Take specialist advice when selecting the right size stove for your room. Often people want to install a stove that is too large, with the result that the stove has to be throttled back, making it run inefficiently and generating more soot and ash than a smaller stove running at full efficiency would do.

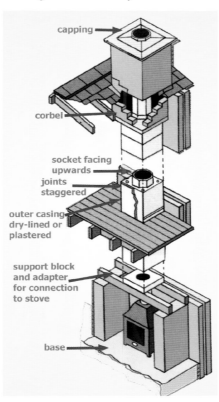

This DM (Double Module) Isokern Chimney System is a precast system made from highly insulating pumice. (Courtesy Schiedel Isokern)

HOW CHIMNEYS WORK

Jonathan Greenall, the man behind Clearview Stoves says: "You can run an ordinary stove on a good chimney but not a good stove on a bad chimney." That's how important it is that a chimney works properly, and to do this, you first need to know what makes a chimney do its job. In other words, why does the smoke go up the chimney rather than drift out into the room?

- The upward movement of air and smoke in a chimney is known as the draught, and we talk about how well a chimney 'draws.'
- A draught occurs because hot gases are less dense than cold gases. Heated gases in a chimney are lighter than the air in the atmosphere, and are therefore drawn up into the chimney.
- All other things being equal, a taller chimney produces more draught because of the simple fact that there tends to be more of this hotter, lighter gas in it, which creates more updraught.
- It's rare that all other things are equal, though! The increased draught inherent in a taller chimney tends to be offset by the greater friction found in the chimney, and the tendency of gases to lose heat as they rise. A very dirty, roughly-lined, tall chimney, subject to poor insulation and possible air leaks, could well be less efficient than a well-designed shorter chimney. Another fact of physics has an effect – albeit a lesser one – on chimney performance. As air is blown over the end of a tube, a drop in pressure is created at the end of the tube. This principle was first discovered by Italian G B Venturi almost 200 years ago, and is the principle by which most carburettors inject petrol into the air stream. Wind blowing over a chimney has a similar effect, adding to draught. Wind eddying down the chimney has the opposite effect, of course.

When a stove and flue are operating really efficiently, the heat from the fire drives gases from the wood which are burned off above the wood in this typically beautiful pattern. The down-wash from the upper air source on a Clearview stove is designed to burn these gases, while at the same time preventing the door glass from sooting up

Michael Oliver of Clearview Chimneys (a separate but associated company to Clearview Stoves) came to fit our new stove and the new flue system.

After the stove had been fitted, I noticed that some of the plaster behind it was loose. You might ask why I didn't replace it before the stove was fitted, and the reason is – I hadn't noticed it then. But, in fact, to replace it then would have been a mistake. The water that unavoidably dribbles down the chimney from the Leca mix will at best stain and might

After clearing away our old wood burner and opening up the chimney ...

... Michael found what he termed a 'moderate' coating of soot and tar on the inside of the stone-built chimney. The chimney was cleaned thoroughly before going any further.

Your installer could use scaffolding to access the top of a chimney like this, but Michael prefers to hire an hydraulic hoist from a local tool-hire company. This is a lot quicker, cheaper, and more versatile than having scaffolding erected for just two days' work, he feels.

The old pot was removed from the top of the chimney ...

... and once protective cloths had been fitted in the living room beneath, Michael and Co were ready to begin scraping tar deposits from the chimney, working from the top down.

Then, working from the bottom up, Michael got ready to sweep and scrape all the remaining deposits he could reach.

This is the Isokern (pumice) lower support block for the flue liner.

Michael made an angle-iron frame to fit into the chimney breast. The two longitudinal members were socketed into the stonework at each end, while the cross-members were welded in place once the exact location of the liner had been established.

With all of the steelwork in place – though not finally welded – Michael slotted the pumice support block for the flue liner onto the frame. He then welded together all of the steel support frame.

For demonstration purposes, these are typically the components you will need. Added to those being fitted by Michael at this stage are the stainless steel components to go inside the room.

The pumice liner, made by Isokern, is light enough to be lowered from above down the chimney a section at a time. Pumice is an ideal material for use as a flue liner, partly because of its relatively light weight, but also because, having been formed in a volcano, it really isn't going to be troubled by the heat from your wood burning stove, is it?

Michael cut a pair of opposing slots in the bottom of the first section of liner to be inserted, so that it could be passed down the chimney on a rope without slipping off. Each separate section would then be added, one at a time from above, until the support at the bottom of the chimney was reached.

This is where the hoist came into its own. At the top of the chimney, the first section of liner was attached to the rope, and lowered part-way down the chimney, ready for the second section to be added to it.

As each section is slotted onto the next, it is held in place with a galvanised steel collar, and the joint is bonded with special fire clay.

When fitting a liner, joints should have their male spigots facing down (though, of course, the fire clay is added with the spigot facing upward). The idea is not to ensure that all the smoke goes up the chimney (it will if joints are properly sealed), but that water running down travels directly to the bottom of the flue, where it will be boiled away again.

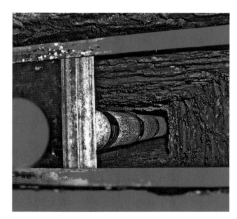

It is most important that the flue should be as straight as possible, and centralised in the chimney. Kinks and bends add to friction in the chimney, which slows the exit of smoke, and if the liner touches the side of the chimney, there won't be any insulation at that point. This tends to be a problem with the inflatable 'sausage' method of lining a chimney.

71

Graham, working with Michael, mixed the Leca insulating material – very lightweight granules made from pumice – with a little cement and a little water.

Leca is light enough to be hoisted roofward in bins before being poured into the chimney.

Back at ground level, Michael pushed the new stove into position. Note the small amount of water dribbling down the plaster: more on that later.

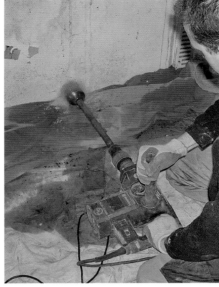

While Graham was up in the air, Michael had been busy cutting a hole through the back wall …

… to take the air inlet pipe for feeding air to the stove from outside the house: this is an optional extra on Clearview's stoves. Perhaps it was the memory of our ancient, tree-eating open fire of decades ago, trying to suck the room and all its contents up the chimney, that decided us to specify the external air supply. Or perhaps it was the modern, well-sealed doors and windows that we've subsequently fitted, which necessitate a guaranteed, adequate air supply. There is certainly now less draught scudding across the floor than we previously experienced. And the dogs and cats, when in wood burner worshipping mode, are extremely grateful …

Stainless steel flue liner is far more common than the Isokern-type we selected because it is much quicker, easier, and cheaper to install. However, its life expectancy is measured in years rather than decades, and its insulating qualities are somewhere between poor and non-existent. Consequently, wood burning stoves fitted with these flue liners are less efficient, won't operate at low settings, and are far more prone to blow-back.

You have to use twinwall (ie insulated) stainless steel to connect the stove to the flue liner, of course. Michael selected from a range of angles and lengths until the right combination was in place.

This liner is telescopic, so can be pushed tightly into place. It was typical of the thoroughness brought to the installation that Michael drilled and tapped the liner where it joined the stove before fitting machine screws, rather than the hit-and-miss self-tapping screws that a less meticulous installer might have used.

Two new courses of engineering bricks took the chimney back to the height I remember it being when we first bought the place …

… and now, with new flashing to protect the top of the chimney from weathering.

The stainless steel bird cover has made for some very noisily disappointed rooks, but at least the chimney will stay twig-free from now on.

There was then just the matter of pointing the air intake hole and fitting a grille to the outside of the wall, behind the wood burner.

Michael explained to Shan how to get the best from the stove. His instructions were spot-on, and you can turn flames up and down almost as fast as those on a gas fire.

even damage any plaster there, so repairs are best carried out afterwards. Anyway, loose plaster operates on the iceberg principle – you only see the most visible 10% until you start looking beneath the surface …

In use, our new Clearview stove is even better than expected. It produces very little smoke (except when first lit), and certainly the door glass remains almost completely 'clear-view' for days on end. When it's running at its most efficient, the heated wood gives off gases that are burned in ethereal swirls of flame, fascinating to watch and efficient in use.

HOW TO LIGHT A WOOD BURNING STOVE

There's no single way of lighting a wood burning stove, but the principle of starting with something that is very easy to light, which is used to ignite kindling before moving on to firewood, is universal.

Whether you decide to use regular or organic firelighters, or rolled up newspaper, the principle is pretty much the same. You will need kindling – thin, very dry strips of wood – and something such as firelighters or newspaper to start off the kindling.

Open the air controls on the stove to about halfway, and open the stove door. Generally speaking, it's best to lay your fire on a thin bed of (cooled) wood ash left from the previous fire.

Place the kindling on the firelighter or newspaper, and light this. As soon as the kindling takes fire, place two or three smaller pieces of firewood from your log box carefully onto the kindling so as not to smother the flames, and close the stove door. When the smaller pieces of firewood have begun burning well, the amount of smoke being created should diminish greatly, and you can begin to add firewood in the usual way.

There's no doubt that, over time, you will develop your own preferences and techniques that best suit your stove, chimney, and combustible materials. However, whatever you do, don't use flammable liquids to start your fire because the risk of an explosion will be very high indeed.

CHIMNEYS AND FLUE LINERS – SUMMING-UP

A flue needs to be warm to be efficient. A cold flue produces colder, denser, and therefore slower-moving, gases, and has a greater risk of blow-back. An unlined or badly insulated flue will 'run' cold and, to make things worse, won't keep gases above condensation point. At two-in-the-morning, gases at the chimney top will be below 100 degrees C, allowing condensation to run down the inside of the chimney, exacerbating the cooling effect, and potentially causing staining.

The quickest, cheapest and easiest type of flue liner to fit is a single-skin, flexible, stainless-steel type ... unfortunately, it's also the worst! It is uninsulated, and may last no longer than a few years. Rigid lengths of insulated stainless flue are preferable, but far better is a rigid, sectional, pumice-type item such as the Isokern, with a semi-dry insulating material called Leca poured around it. Wet-poured liner (formed around a 'sausage' previously inserted into the chimney) can be prone to problems. The 'sausage' can touch chimney walls, meaning there's no insulation at that point.

Note that a stove can run slower with an insulated liner because it doesn't need to create (and waste) as much effort maintaining heat in the chimney.

The best way to install a new flue in, say, a chapel conversion is to build a central fireplace and a flue right through the centre of the house, bearing in mind building regulations with regard to surrounding flammable materials. The flue stays warmer, heat is more efficiently used throughout the home, and the top of the flue is nearer to the ridge.

- Keep a stand-alone flue as straight as possible. Every bend introduces friction, and sharp bends are worst of all.
- Even an insulated stand-alone flue may smoke horribly in the middle of winter if on the outside of a building.

• TIP: If you're building a new house with an open fireplace, don't fit a concrete lintel too far back in the flue opening, because this makes it impossible to fit an insulated flue liner later. Also, beware a builder who wants to fit a square liner: this will touch chimney bricks all-round and cannot be properly insulated.

Cleaning frequency

Slow-moving gas, like water in a meandering stream, will deposit solids on the inside of the chimney. If you see smoke rolling picturesquely out the top of your chimney it's too slow! A warmed flue liner improves momentum.

The chimneys in Clearview Stoves' showroom are cleaned about every ten years but some chimneys need sweeping every couple of years. Poor combustion can mean fifty times more smoke going up the chimney – and unburned creosote and soot will stick to the sides of the chimney.

Smoke in the room

Usually, this is caused by a blocked or low chimney (such as on the side of a barn conversion where the outlet is below the ridge). Clearview Stoves claim it's rare to encounter blow-back where the chimney outlet is 600mm (23ins) or more above the ridge.

• TIP: If building your own chimney, look at what works for the surrounding properties, which will have been built or adapted for the environment.

To prevent birds nesting in your chimney top, you may need a bird cover. Don't use chicken wire – it's awful and will collect soot and eventually seal the top of the chimney more effectively than a nest would! Make sure that the bird cover contains enough room for a flue brush to appear so that your sweep can check he has swept to the chimney top.

• TIP: In unavoidably adverse conditions, such as with a chimney on a low building adjacent to a taller one, you may need to resort to an electric chimney fan. If unsure whether or not one will be required when building, at least have the wiring put in place

Insufficient air coming into the room can also cause blow-back. Bear in mind that, traditionally, houses had plenty of through-draughts, whereas today they are often too airtight for an open fire. If you don't provide sufficient oxygen for the fire, it will find its own, pulling great gulps of air down the chimney and belching smoke out into the room. If necessary, add a vent to an outside wall adjacent to your stove or fireplace.

PART 4: FITTING A BACK BOILER

When choosing a wood burning stove, consult a specialist to help you select the right model. Almost no-one choses a stove that's too small, but an oversized stove is a big mistake. Running efficiently, it will make the room too hot; turned way down, it won't be efficient and may regularly smoke-up. However, when you want to add a backboiler to heat domestic hot water, all those calculations will need to be done again because water heating sucks a significant amount of heat away from a room heating stove, and so, to allow for the water heating side of things, the stove will need to have extra size and extra capacity – and will, of course, burn more wood.

You will need to assess the feasibility and practicability for yourself, in association with your specialist. However, if you can install a back boiler – and if this is properly designed and installed – you'll find yourself with lashings of hot water all

All wood burning stoves designed for the optional extra of a back boiler should have a pair of holes partly stamped out of the back. After removing the fire brick and locating the position of the stamping ...

... the stamped-out piece of steel can usually be knocked out with a hammer. In a few cases, it may be necessary to drill the body of the stove.

The back boiler connections should be sealed where they pass through the stove, so it's necessary to make four loops of fire rope, connecting the butt ends with fire clay.

Two of these loops have been pushed on to the threaded connectors on the back boiler by fitter Justin, with a wipe of fire clay on each side of the loop ...

... before he inserted the purpose-made stainless steel back boiler into the stove, and passed the threaded connectors ...

... through the holes in the casing of the stove, before sliding on another loop of fire rope.

Justin then added a large, stainless steel washer and nut to each of the connectors, carefully tightening them evenly.

Once in place, the back boiler takes up very little extra room than did the fire brick and once it's got some soot on it, it won't be noticeable. The connectors on the rear are used, of course, for connecting the pipework that will provide hot water ...

... to the hot water cylinder. As you can see, this is a much larger hot water cylinder than is used in conventional domestic installations. It was specially produced by Newark Copper Cylinders to be suitable for hot water supplied by several different renewable sources.

We chose to use 28mm copper pipe because the circulation of hot water from a back boiler usually depends on the process of convection rather than using a pump, and the larger the pipe (up to 28mm), the smaller the frictional losses. Generally speaking, circulating the water by convection is a much safer approach than using an electric pump because overheating caused by pump failure – with its associated risk of an explosion – is precluded. Convection takes place when water is heated, which makes it expand. It thus becomes less dense than the cold water in the pipe and rises to the hot water cylinder, where it gives up its heat into the cylinder and, because it's now cooler and denser again, returns to the back boiler to be reheated.

75

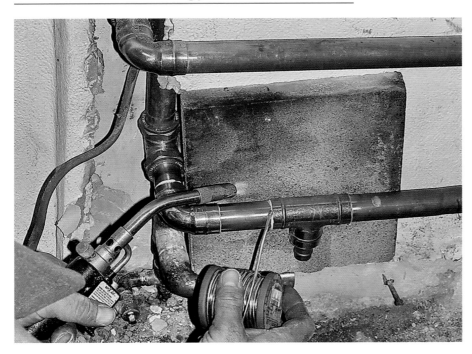

through the winter at a time when the solar domestic water heating system is at its least effective.

A word of warning, however: the water in the hot water cylinder can easily become so hot that it becomes unsafe for children or elderly people to use. In these cases, a mixer valve installed near to the hot water cylinder will regulate the output temperature by adding measured amounts of cold water in order to bring it down to a safe level.

Where Justin made soldered connections downstairs, in the fireplace, he used a heatproof mat to protect the wall behind.

The need for extensive pipework in some locations can make the installation of a back boiler a costly business. Fortunately, we had installed the main pipe runs many years ago when we first purchased and developed the property.

Chapter 6
Solar PV electricity generation

This chapter contains more information on design and theory than any of the others in this manual. That's because, in order to know what you can fit and where you can fit it, you must be aware of the factors that influence the efficiency of photovoltaic (PV) panels, and also the strict regulations surrounding them in most countries.

For instance, if the only place that you can fit panels is in a direction within the arc between north-east and north-west, they almost certainly wouldn't be worth installing in the northern hemisphere. Similarly, if the only place you can fit them is completely shrouded by trees; again, they wouldn't generate much electricity. And between those two sets of extremes, there are many other factors to bear in mind when deciding whether and where to fit PV panels; how to install them, and how many to install.

On the other hand, many technical factors to do with the specification of PV panels, inverters and associated components are the domain of systems designers, and mere mortals like us don't need to worry about them. Many combinations of PV equipment have been proven in practice, and

this makes it relatively easy to choose an off-the-peg system, or a variation of one that will suit your particular needs. Some of the following advice and illustrations have been supplied by leading UK PV installer Eco2Solar: a company noted for its attention to detail. But first of all, even before calling in the specialists, it helps to know what is possible …

SECTION 1: PLANNING

In the UK, only components, equipment and assemblies that have been approved by MCS should be used. This might seem overly-bureaucratic but, in fact, it is of tremendous benefit to know that all of the components that you can choose from are of a good standard, thus avoiding the risk of purchasing substandard, unsafe, inefficient or poorly produced equipment. Details of MCS-approved system components can be found at www. microgenerationcertification.org.

The DC system

Part of the following introduction is based on advice from the UK's Energy Saving Trust (EST).

Orientation and tilt

There are some things you can't easily change, such as the direction your home is facing! Your roof should ideally face due south if you are in the northern hemisphere, or true north if you are in the southern hemisphere. In the UK, it would have a pitched angle of around 30° from the horizontal to give the best overall annual performance. Installations at any pitch and facing anywhere to the south of due east and due west are feasible, although output and income will be reduced. Installation is not recommended on roofs facing north, which applies to some of the unfortunate homeowners surrounding this Eco2Solar installation. The author's roof, seen in more detail later, faces just south of south-west – not perfect, but perfectly acceptable.

Incidentally, if you're trying to be 'spot-on' (and why not!), a warning from Californian specialist Charles Landau: "True north is not the same as magnetic north. If you are using a compass to orient your panels, you need to correct for the difference, which varies from place to place. Search the web for 'magnetic declination' to find the correction for your location."

On the other hand, as long as your roof is within an acceptable arc, you can still fit panels that will generate a very useful amount of electricity. This installation makes the most of two roofs whose angle is intersected by the 'perfect' south-facing angle.

If your roof is completely unsuitable and you have sufficient space, you might want to consider an array of PV panels on a frame mounted on the ground. This has the advantage that installation costs will be lower; the angle can be adjusted to the optimum, and you might even be able to vary the tilt angle – see later. But whichever you choose, this Eco2Solar picture is a great reminder that it's the earth's future that is the biggest consideration.

There are detailed instructions later in this chapter on measuring shadow fall, but you can do most of what you need to using a watch, a diary, and a dose of common sense. This photograph was taken on May 29 in central England, and shows clearly where the morning shadows from the chimney and roof lay their cloaks. This is something that, ideally, should be measured and checked at several times of the day, and at regular intervals throughout the year. The problem is that a small amount of shadow can have a disproportionately large affect on PV panel output.

• *TIP! The site www.gaisma.com is an invaluable resource for carrying out detailed calculations. It provides, for a vast range of towns and cities around the world, the correct sunrise, sunset and solar insolation data.*

This is another roof on which panels were to be placed. Shadowing earlier and later in the day is least harmful because that's when there is least energy available from the sun, so there is less to lose. Most installations require some compromise.

TILT degrees	West									South	East								
	90	80	70	60	50	40	30	20	10	0	10	20	30	40	50	60	70	80	90
0	87	88	90	91	92	92	93	93	93	93	93	93	92	92	91	90	89	87	86
10	84	87	90	92	94	95	95	96	96	97	97	96	95	94	93	91	89	87	84
20	82	85	90	93	94	96	97	98	99	99	98	97	96	95	93	91	88	84	81
30	78	83	87	91	93	96	97	98	99	100	98	97	96	95	93	89	85	81	78
40	75	79	84	87	92	94	95	96	96	96	96	95	94	92	90	86	82	77	72
50	70	74	79	83	87	90	91	93	94	94	94	93	91	88	83	80	76	73	70
60	65	69	73	77	80	83	86	87	87	87	88	87	85	82	78	74	71	67	63
70	59	63	66	70	72	75	78	79	79	79	79	79	78	75	72	68	64	61	56
80	50	56	60	64	66	68	69	70	71	72	72	71	70	67	66	60	57	54	50
90	41	49	54	58	59	60	61	61	63	65	65	63	62	59	60	52	50	47	44

This table from the Energy Saving Trust (EST) shows the percentage variance in performance when orientation and tilt are adjusted away from the optimum (100%). What might be surprising is the fact that, in northern Europe, a panel with zero tilt can still produce 90% of the theoretical maximum when pointing a full 90 degrees away from due south. In other words, even when the only available angle of installation means you can't achieve maximum potential output, you may still be able to generate a significant and useful amount of electricity.

At the other end of the scale, for those with commercial or institutional buildings with flat roofs or an appropriate orientation, the sky is almost the limit when it comes to PV panel installations, as Eco2Solar's MD, Paul Smith, proudly demonstrates.

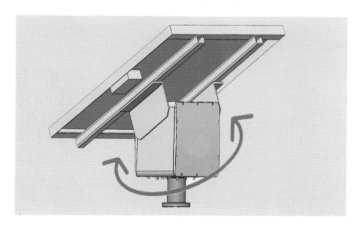

Another solution that might seem a touch 'off the wall,' but which can bring about a significant yield where space is limited, is to use a solar tracker. This follows the angle of the sun as it 'moves' through the day, maximising the amount of energy absorbed, albeit at the expense of some relatively costly equipment. Some systems track horizontally while others can also make vertical adjustments to allow for varying solar angles at different times of the year.

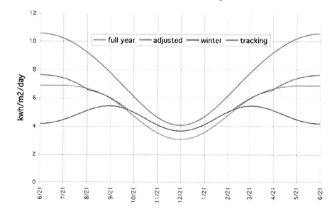

This drawing from Charles Landau, graphically illustrating the results of his performance measurements taken in southern California (see later in this chapter), demonstrates some of the differences between fully tracked performance and panels that have had their positions (manually) adjusted at different times of the year.

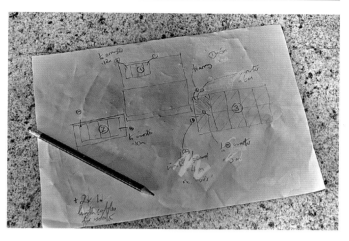

Having decided where your panels and all the rest of the hardware is going to be located, you'll need to have a sketch plan in order to begin the formal planning process. Usually, this stage makes you realise why you can't do some of the things you might have wanted to …

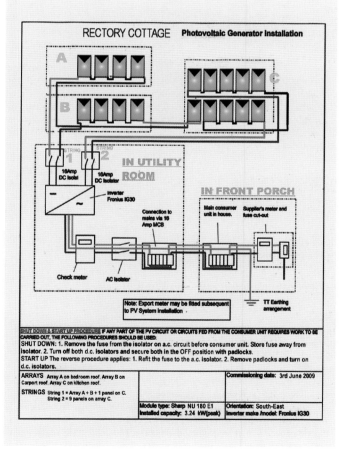

After the work has been completed, you will need to have an effective plan of the layout on display, such as the one shown here. Installers find it very worthwhile to have an interim version of this plan drawn up before installation begins.

Detailed planning

You, or more likely your installer, will next have to address detail design issues, now that you've established what is and is not possible with the layout you have available, and these will almost certainly include revisions to the initial plans as the detailed design calculations are carried out, but that's just a normal part of any design process. So, the following are those technical details that will need to be considered during the detailed design process.

The following information, up to the heading *Solar PV tilt effect*, is based on information supplied by the Energy Saving Trust (EST), and provides further details about what needs to be considered in the PV design process.

Solar electricity generation doesn't necessarily require direct sunlight: as much as a third of the energy generated on a sunny day is possible on a cloudy day at the same time of year ...

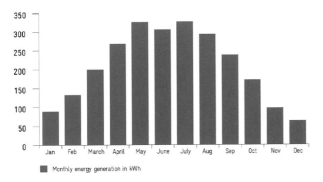

Monthly energy generation in kWh

This chart from the EST shows a typical seasonal spread of energy generation for a south-facing system of 3kWp. Not surprisingly, significantly less electricity is generated during winter months than during the summer months.

... although a complete covering of snow does put a stop to proceedings!

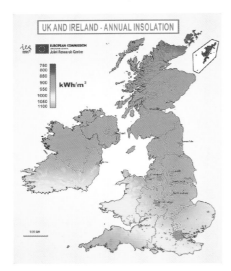

UK AND IRELAND - ANNUAL INSOLATION

The amount of electricity generated by a solar PV system can also vary depending on where you live, as this map produced by the European Commission demonstrates. Northern areas (in the southern hemisphere) receive less energy from the sun over the year than southern areas. For example, a 1kWp system will generate less electricity in northern Scotland than it would in Cornwall. However, solar electricity generation is still worthwhile; the differences aren't substantial. You can get an estimate of how much a system will generate in your location (and how much it will earn) using various online solar energy calculators.

If you live in a region where you regularly receive snowfall, part of your calculation might include trying to ensure the panels are mounted at a steeper pitch so that snow easily slides off the fairly slippery surface of the panels, which will warm up in any case as soon as they begin to generate electricity. This is what has happened with this Eco2Solar installation.

Shading

All of the modules are connected, so any shading on a single module will unavoidably affect the performance of the entire array. A system can tolerate some shading early or late in the day without much reduction in overall output, but it should not be shaded between 10am and 4pm. Nearby trees, chimneys, TV aerials and vent pipes are all common causes of shading, and should be accounted for before any installation.

Planning permission

In England, Wales and Scotland, planning permission is not required for most home solar electricity systems, as long as they're below a certain size, but check with your local planning officer, especially if your home is a listed building, or in a conservation area or World Heritage site. We can only give general guidance here: you'll find full details on the relevant government's legislation websites.

Permitted development rights for solar PV (roof-mounted) – UK except Scotland –
Permitted unless panels protrude more than 200mm (8in) when installed. Your local Building Control Office may want to check that your roof structure is suitable – your installer should be able to advise on this.
Scotland: permitted unless –

- Installation is on any part of the external walls of the building if the building contains a flat.
- When installed on a flat roof the panels are situated within one metre of the edge of the roof, or protrude more than one metre above the plane of the roof.
- When installed panels project higher than the highest point of the roof (excluding the chimney).
- The building is within a conservation area or World Heritage site, and the solar PV or solar thermal equipment is installed on a roof which forms the front of the building and is visible from the road.
- Plus: the solar PV equipment must, as far as is reasonably practicable, have minimum effect on the amenity of the area, and be removed when it is no longer needed or used for domestic microgeneration.

Building regulations

All installation work in and around occupied structures in the UK are covered by building regulations. You should check for local equivalents of outside the UK. Though these differences are of a relatively minor nature, you will find that your Local Authority Building Control (LABC) will have readily accessible information on what is and is not allowed, and what permissions you may need to apply for. The two principle methods are –
1. Submitting a building notice to LABC before work starts.
2. Notification via a Competent Person Scheme* – can be after the work has been completed. (Most people employ contractors who use method 2 because method 1 can be expensive and time-consuming.)

It is especially important when structural work is being carried out – such as to alter or strengthen a roof before fitting the PV system – that prior agreement is reached with your LABC in the UK. However, where it has been determined (ie calculated) that the structure can accept the loads concerned, either a building notice or notification via the Competent Person Scheme* meets building regulation requirements.

*The Competent Person Scheme allows previously registered businesses and individuals to self-certify work that complies with UK building regulations – as an alternative to submitting a building notice or incurring the fees of a local authority-approved inspector. Those who are not registered with a self-certification scheme – including DIY-ers – need to notify and/or submit plans to LABC, unless the work is non-notifiable.

Advice about issues that are specifically relevant to Scotland can be found on the Local Authority Building Standards Scotland website: www.sabsm.co.uk

The useful website, yougen.co.uk, makes the additional point that: 'Even though your local planning office will probably give you the green light ... this does not automatically mean your roof is good to take the weight of added load.' It goes on to say that a surveyor will need to inspect the roof structure, be this a qualified* surveyor supplied by your chosen installer, or one you select yourself. (*Do ask to check their accreditation – a trustworthy installer shouldn't mind this at all.)

House insurance

Many buildings insurers will supply cover under the terms of your current policy if you have solar PV panels installed on the roof of your home. But always contact your current insurance provider for advice before having solar PV panels installed, because –

- The installation of solar PV panels is a material change to your home's structure that your insurance provider needs to know about.
- You must know exactly what your insurance covers: does it cover damage caused by the installation; does it cover theft?
- Some types of installations, such as ground-mounted systems, may not be covered as standard.

As a precaution, the EST recommends you confirm in writing with your insurance provider that you have had a solar PV system installed, and ask it to confirm in writing that this has been received; also the terms of the cover being provided. If you purchase a home with a solar PV system already installed, or change insurance provider, make sure it is aware of the system before taking up insurance.

Continued on page 84

SOLAR PV – THE TILT EFFECT

Something that is obviously not applicable to installations on roofs where there is almost never an opportunity to adjust the tilt angle, the information in the next paragraph about tilt effect is reproduced with the kind permission of Charles Landau in California, and is the result of some of his own research and calculations. It shows how complex the establishment of 'ideal' tilt angles can be, and indicates how much the perfect tilt angle varies for time of year, time of day, latitude, and even which hemisphere you're in. In practice, inverter manufacturers' online software generally carries out all the calculations you will need for a fixed installation and, theoretically, could be used to show the effects of changing the tilt angle, when that is possible.

This advice applies to any type of panel that derives energy from the sun: photovoltaic, solar hot water, etc. We assume that the panel is fixed, or has a tilt that can be adjusted seasonally. (Panels that track the movement of the sun throughout the day can receive 10% (in winter) to 40% (in summer) more energy than fixed panels. This section doesn't discuss tracking panels.) And it's worth passing on some words of wisdom from the creator of this section, Charles Landau, who says: "Don't obsess about the exact angles just because I've calculated them to the tenth of a degree. A difference of a few degrees will make very little difference to the energy you gather."

Tilt angles explained

A zero tilt angle means that the face of the panel is aimed directly upward. A positive tilt angle means that the panel faces toward the equator more. In the northern hemisphere this would mean tilting it so that it is south-facing. On rare occasions, the tilt angle can be negative, which means the panel faces away from the equator.

Fixed or adjustable?

It is simplest to mount your solar panels at a fixed tilt and just leave them there. But because the sun is higher in the summer and lower in the winter, you can capture more energy during the year by adjusting panel tilt according to season. The following table shows the effect of adjusting the angle, using a system at 40°N latitude as an example.

	Fixed	Adjust 2 seasons	Adjust 4 seasons	2-axis tracker
% of optimum	71.1%	75.2%	75.7%	100%

So, at 40°N, adjusting the tilt angle twice a year will provide a meaningful boost in energy. Adjusting four times a year produces only a little more, but could be important if you need to optimize production in spring and fall (autumn).

The graph on page 79 demonstrates the effect of adjusting the tilt. The turquoise line shows the amount of solar energy generated each day if the panel is fixed at the full year angle: the red line shows how much would be generated by adjusting the tilt four times a year as described below; for comparison, the green line shows the energy generated from two-axis tracking, which always points the panel directly at the sun. (The violet line is the solar energy per day if the panel is fixed at the winter angle, discussed below.) These figures are calculated for 40° latitude.

Fixed tilt

A fixed angle is convenient but does have some disadvantages –
1. Less power will be generated than if the angle was adjusted, and
2. If snow is common in winter, adjusting the panels to a steeper angle then makes it more likely that they will shed snow (a snow-coveredpanel produces little or no power). Use one of these formulae to find the best angle from the horizontal at which the panel should be tilted –
• If your latitude is below 25°, use the latitude x 0.87.
• If your latitude is between 25° and 50°, use the latitude, x 0.76, + 3.1 degrees.
• If your latitude is above 50°, see 'Other situations' below.

Adjusting the tilt angle twice a year? The following table gives the best dates on which to do this –

	Northern hemisphere	Southern hemisphere
Adjust to summer angle on	March 30	September 29
Adjust to winter angle on	September 12	March 14

If your latitude is between 25° and 50°, the best tilt angle for summer is the latitude, times 0.93, minus 21 degrees. The best tilt angle for winter is the latitude, times 0.875, plus 19.2 degrees. If your latitude is outside this range, see 'Other situations' below.

Adjusting the tilt four times a year? This might be best for you if connected to the grid, and can use or sell all the power you produce. The following table gives the best dates on which to adjust –

	Northern hemisphere	Southern hemisphere
Adjust to summer angle on	April 18	October 18
Adjust to autumn angle on	August 24	February 23
Adjust to winter angle on	October 7	April 8
Adjust to spring angle on	March 5	September 4

If your latitude is between 25° and 50°, the best tilt angles are –
- For summer, take the latitude, multiply by 0.92 and subtract 24.3 degrees.
- For spring and autumn, take the latitude, multiply by 0.98 and subtract 2.3 degrees.
- For winter, take the latitude, multiply by 0.89 and add 24 degrees.

If your latitude is outside this range, see 'Other situations' to the right.

In winter, a panel fixed at the winter angle will be relatively efficient, capturing 81 to 88 percent of the energy compared to optimum tracking. In the spring, summer, and autumn, efficiency is lower (74-75% in spring/autumn, and 68-74% in summer), because in these seasons the sun traverses a larger area of the sky, and a fixed panel can't capture as much of it. These are the seasons in which tracking systems give the most benefit.

Note that the winter angle is about 5° steeper than commonly recommended because, in the winter, most of the solar energy comes at midday, so the panel should be pointed almost directly at the sun at noon. The angle is fine-tuned to gather the most total energy throughout the day. Summer angles are about 12 degrees flatter than is usually recommended: in fact, at 25° latitude in summer, the panel should actually be tilted slightly away from the equator.

Tilt fixed at winter angle
If your need for energy is highest in the winter, or is the same throughout the year, you might just want leave the tilt at the winter setting (this could be the case if, for instance, you are using passive solar to heat a greenhouse). Although you could get more energy during other seasons by adjusting the tilt, you will get sufficient energy without doing so.

Time-of-use rates
In some grid-connected systems, energy is more valuable during peak periods, and this may affect your calculations. There is more about this on the www.solarpaneltilt.com website.

Other situations
Some situations are more complex than can be accommodated by a simple formula. Specialists, such as Charles Landau, can carry out calculations for a consulting fee: see www.solarpaneltilt.com.

Background to the calculations: assumptions
These calculations are based on an idealised situation, and assume that you have an unobstructed view of the sky, with no trees, hills, clouds, dust, or haze ever blocking the sun. The calculations also assume that you are near sea level. At very high altitude, the optimum angle could be a little different.

Why does this work?
The recommended angles can seem counterintuitive. For example, consider summer at 40° latitude. At noon on the solstice, the sun is 40°-23.5°, which is 16.5° from directly overhead. To capture the most sun at that time you would tilt the panel 16.5° to point it directly at the sun. On other days during summer, the sun is a little lower in the sky, so it would be necessary to tilt the panel a little more, yet the advice is to tilt it only 12.5°. The sun is never that high: how can this be right?

The answer is that we are considering the entire day, not just at noon. In the morning and evening, the sun moves lower in the sky, and also further north (if in the northern hemisphere). It is necessary to tilt less to the south (or more to the north) to collect that sunlight.

How these figures were calculated
Percentages may not be exact due to rounding up or down. For each configuration of latitude and season, over 12,000 data points were calculated for various times throughout the day and the year. For each data point, the equations of celestial mechanics were used to determine the height and azimuth of the sun. The intensity of the sun was corrected to account for the increased absorption by the atmosphere when the sun is lower in the sky, using the formula –

Intensity in kW/m^2 = 1.35x(1.00/1.35) sec (angle of sun from zenith)

This formula assumes that the earth is flat, so a factor was applied to account for the curvature of the earth (and therefore the earth's atmosphere). These factors, and the angle of the sun with respect to the panel, then determine the insolation on the panel. An iterative method then determined the angles that give the maximum total insolation each season. Given those angles, the beginning and ending dates of the season were adjusted to the optimum, then the angles recalculated until the process converged. After the optimum dates and angles were calculated, it was determined that a linear formula closely approximates the optimum. The formulae are only accurate within the specified range.

PART 1: PLANNING THE INSTALLATION

Huge amounts of technical detail and required (UK) specifications are available from Microgeneration Certification Scheme (MCS): full details can be seen on the MCS website: http://www.microgenerationcertification.org/.

Locations

Accessories and equipment must be located so that any future servicing and maintenance can be carried out efficiently.

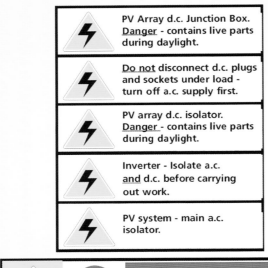

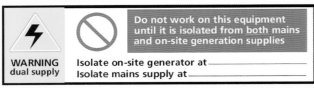

All labels must be clear, legible, located so as to be easily visible, durably constructed and affixed to last the lifetime of the installation. Labels are readily available from online suppliers, although not all sell labels that are 1. Compliant with regulations regarding typeface, sizing, etc, 2. Correct for current installations, 3. A complete set. Check online for current requirements; purchasing from an established installer should ensure you obtain the correct types and quantities of labels. (MCS)

Labels

Cables

Selecting the correct cables to use is straightforward, provided that correctly specified, purpose-made, PV cables are used with the correct terminals.

• Cable runs should always be designed to be as short as possible because voltage drops on DC systems are much more pronounced than on higher voltage AC systems. Plug-and-socket connectors simplify and enhance the safety of the installation, and are recommended in particular for any installation carried out by a non-PV

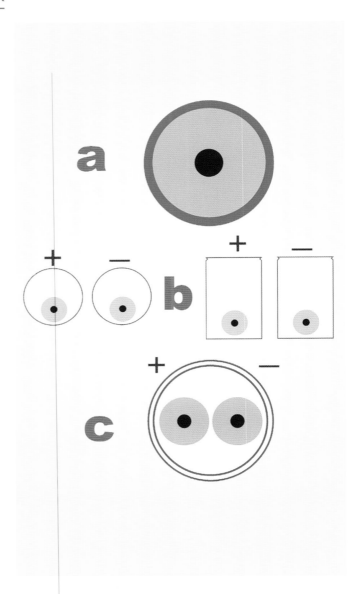

Examples of sections through MCS-approved cables –
a) Single conductor, double-insulated cable. b) Single conductor cable suitably mechanically protected conduit/trunking. (Alternatively, a single core SWA cable may be a suitable mechanically robust solution). c) Multi-core Steel Wire Armoured (SWA). Typically only suitable for main DC cable between a PV array junction box and inverter position, due to termination difficulties between SWA and the plug and socket arrangements pre-fitted to modules.

specialist: eg a PV array being installed by a roofer.

PV array cables should not be buried in walls or otherwise hidden in the building structure as mechanical damage would be very difficult to detect, and may lead to an increased risk of electric shock and fire risk. Where cables have to be buried or hidden, they must be suitably protected from mechanical damage with: metallic trunking, conduit or steel wire-armoured cable in accordance with local regulations.

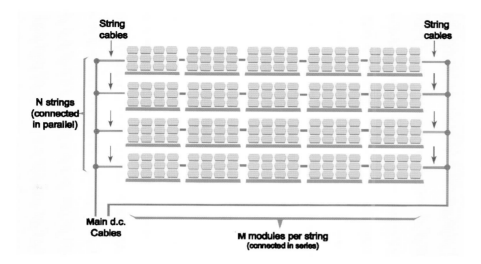

STRING CABLES SAFETY. This area is mentioned here for general interest only because it is something that an experienced PV system designer should automatically be aware of –
• In a PV array with a number of strings, a fault could cause currents to flow though parts of the DC system, posing a considerable fire risk.
• For small systems, the string cables are suitably oversized so that any fault current can be safely accommodated, precluding a fire risk from overloaded cables. (MCS)

Other in-line cable connections
In general cable junctions must either be via an approved plug and socket connector or contained within a DC junction box. Soldering should be avoided unless essential: eg a soldered extension to a module flying lead.

Dc plug and socket connectors (MCS).
• Cable connectors must not be used as a method of DC switching or isolation under load because DC arcing (which is much more pronounced than arcing in AC systems) can cause permanent damage to connectors.
• Plugs and socket connectors mated together in a PV system must be of the same type, from the same manufacturer, and comply with the requirements of BS EN 50521 or in an MCS test-approved combination (or the other regulations).
• Easily accessible connectors must be of the locking type, requiring a tool or two separate actions to separate, and must have a sign attached that reads: 'Do not disconnect DC plugs and sockets under load.'
• Plug and socket 'Y' connectors can also be used to replace a junction box. It is good practice to keep 'Y' connectors in accessible locations and, where possible, note their location on layout drawings to assist troubleshooting.

DC isolation and switching
There is a requirement for both isolation and switching in the DC side of the PV array circuit. (Note that an additional DC switch or isolating device may be specified for systems with long DC cable runs.)

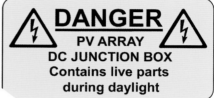

⚠ DANGER ⚠
PV ARRAY
DC JUNCTION BOX
Contains live parts
during daylight

If there is more than one string, they are normally connected in parallel in a DC junction box (or 'combiner box'), labelled as shown. Note: PV panels cannot be turned off: terminals will remain live during daylight hours, and it is important to ensure that anyone opening an enclosure is fully aware of this. (MCS)

The selection and design of MCBs (miniature circuit breakers), string fuses and blocking diodes* are complex and highly technical areas, and should be left to specialists in PV system design.
*Not to be confused with bypass diodes which are normally built into a junction box on the back of each PV module, and are intended to allow current(s) to bypass cells and/or modules that have a high resistance, usually caused by shading.

More specialist detail
An easily accessible load-break switch disconnector, on the DC side of the inverter must be provided, and MCS has much specific data about this. However, at the time of writing, regulations in this area were subject to review. Once more, this whole area of DC switching is something that any qualified PV designer and installer will (or should!) know about. Similarly, there are electrical installation requirements, such as those to do with earthing (grounding) and islanding (see Glossary), insulation, isolation, lightning protection, surge protection measures, and others that are specifically laid out by MCS and which require the services of trained, qualified personnel for both design and installation.

PART 2: THE AC (MAINS) SYSTEM
Notes –
- So that everyone is using the same terminology, the public supply at the point of the AC switch (disconnector), should be regarded as the 'source,' and the PV installation the 'load.' (In other words, for the purpose of making the connections to the mains AC system, the PV panels are NOT regarded as the 'source.')
- It is recommended that Inverters carry a sign stating: 'Inverter – isolate AC and DC before carrying out work.'

AC cabling

MCS states that the PV system inverter(s) should be installed on a dedicated final circuit to the requirements of BS 7671, in which –
- No current-using equipment is connected (although this does not include data loggers).
- No provision is made for the connection of current-using equipment.
- No socket outlets are permitted.

However, it seems that this does not preclude the installation of a hybrid system with battery back-up where, in the event of a power cut, the building's AC system is disconnected from the grid, since some MCS-approved installers are offering just such hybrid systems. This would, presumably, be because the PV system would, with power cut to the house, supply the batteries rather than being directly connected to the 'final circuit.' If interested in this system, check with your MCS-approved supplier.

MCS also specifies regulations regarding protective devices (including RCD) and AC voltage drop limits. Once again these are highly technical matters and are the concern of the designer and installer. In any case, they maybe largely irrelevant because equipment manufacturers which supply off-the-shelf packages should always ensure that the equipment and specifications meet local regulations – but check if buying from abroad!

AC isolation and switching

To comply with requirements, the PV system must be fitted with an isolator (switch) adjacent to the inverter, to disconnect the inverter from the source of supply (AC). This should be to MCS specification and labelled as 'PV system – main AC isolator.'

INVERTERS: The PV system must disconnect when the grid supply is not energised to prevent the photovoltaic system feeding the grid during a loss of mains power. This is termed 'islanding,' and poses a potential danger to those working on the network/distribution system. For that reason (and for other technical reasons), an inverter used in the UK must carry a G83/G59 type test certificate (which is common to all approved inverters), 'unless specifically agreed by an engineer employed by or appointed by the DNO for this purpose, and in writing'.

Inverter sizing

Determining inverter size for a grid-connected PV system is influenced by a number of factors, including the fact that said inverter has been approved for use in the territory in which the system is sited. All of the factors affecting inverter performance will be shown on the website of all reputable inverter manufacturers, including performance figures under different criteria and circumstances. So, in practice, provided that all of the relevant information on the solar panels is to hand, any difficulties in inverter matching are obviated. Typically, each manufacturer has online sizing software which enables the designer (or aspiring designer – which, in this case, could be you, who would then present your preferences to a qualified specialist) to calculate the best and most cost-effective combination of solar panels and inverter(s), bearing in mind the space you have available

for fitting PV panels. This can be a very enjoyable exercise to carry out, as you use the inverter manufacturer's software to 'game plan' as many different permutations as required. If you try to go outside ofmsafe or efficient parameters, the software invariably lets you know! Where issues are not quite black-or-white, most manufacturer's software provides you with percentage efficiency figures so that you can make your choices – and whatever compromises might be necessary.

INVERTER VENTILATION: Inverters generate some heat and need sufficient ventilation, as specified by the manufacturer. Inverter locations such as plant or boiler rooms, or roof spaces prone to high temperatures, should be carefully avoided. The inverter will 'turn itself down' when it reaches maximum operating temperature. Ideally, place a label stating: 'Do not block ventilation' next to the inverter.

So, even though this inverter in the author's second installation is situated in an outdoor shelter and housed in an oak and elm cupboard, it has plenty of space around it, and the top and bottom of the cupboard have been left open to allow air circulation.

METERING. Inverters generally display basic output figures and, although an export meter is not directly part of a PV system, one will be needed in order to enable any payments available for exported electricity. Your energy supplier will usually install such a meter when required. (MCS)

MOUNTING SYSTEMS: There are many different roof fixing types available, dependent on the type of roof covering in use, and it's essential that the correct fixing system is chosen, and installed in accordance with the manufacturer's instructions. All on-roof components and in-roof products (eg PV tiles), plus their fixing and flashing components must be specifically approved to work together and with the named PV module, or be listed as approved universal components. The MCS installer must install to the manufacturer's instructions.

PART 3: ROOF FIXINGS AND SOLAR PANEL MOUNTINGS

The manufacturer's instructions should always be followed precisely when installing the mounting structure of a PV array.

- Permitted clamping zones are prescribed by each manufacturer, and often vary.
- Wind loads vary from site to site, and the designer or installer must ensure that the design wind load is within the specified range – especially important where 'universal' components are used.
- Thermal expansion should be considered when installing larger arrays.

Generally, systems should be kept away from the edges of roofs, leaving a minimum clearance zone of 40-50cm (15-20in), because –

- Wind loads are higher in edge zones.
- Keeping edge zones clear provides better access for maintenance and fire services.
- Taking arrays close to the roof edge may adversely affect rain drainage.

If a PV array mounting system is fitted to an unusual roof, such as one with metal or GRP cladding, the strength of the roof covering and its adequacy to transfer all additional loads to the supporting structure must be verified by a qualified person.

PART 4: ASSESSING POTENTIAL PERFORMANCE
Shade effects

As previously stated (although its importance cannot be overstated) shade has a big impact on the performance of a PV system. Even a small degree of shading on part of an array can make a very significant difference to overall output. Shade is one element of system performance that can be specifically addressed during system design, by careful selection of array location, equipment selection and layout, and in the electrical design (string design to minimise shade effects). Shading from objects adjacent to the array (for example, vent pipes, chimneys, and satellite dishes) can have a very significant impact on system performance. Where such shading is apparent, either position the array out of the shade zone, or, where possible, relocate whatever is casting the shade. Installing a system with significant shading nearby will have a considerable affect on performance. See the following calculations for further information.

Temperature effects

Any increase in PV module temperature results in a decrease in performance (eg 0.5% per 1°C above STC* for a crystalline module). Sufficient ventilation behind an array must be provided for cooling: typically a minimum 25mm (1in) vented air gap to the rear, but it is essential to follow manufacturer instructions. For instance, for at least some Hyundai PV panels, the manufacturer states: 'Space between PV module frames and installation objects is necessary for cooling air circulation. Do not seal this space. Minimum 4 inch (100mm) of standoff height is necessary based on UL Fire Class C.'

*STC stands for 'standard test conditions' specified for PV cells and modules, which are measured at 25°C, with a light intensity of 1kW/m², and an air mass (AM) of 1.5. (AM 1 occurs when the sun is directly overhead on the equator; AM numbers larger than 1 indicate that the sun is at an angle.)

Other factors

A variety of other factors will also affect system performance, including –

- Panel characteristics and manufacturing tolerances.
- Inverter efficiency.
- Inverter – array matching.
- Cable losses.
- Soiling of the array (more relevant in certain locations).
- Grid availability.
- Equipment availability. If the system is down because of equipment failure, it won't be generating electricity or income. Carry out some research into which equipment is most reliable – and most readily available.

MCS METHODOLOGY

MCS standard estimation method –

1. Establish the electrical rating of the PV array in kilowatts peak (kWp).
2. Determine the postcode (region).
3. Determine the array pitch.
4. Determine the array orientation.
5. Look up kWh/kWp (Kk) from the appropriate location-specific table*.
6. Determine the shading factor of the array (SF) according to any objects blocking the horizon – using shade factor procedure (see following pages). The estimated annual electricity generated (AC) in kWh/year of installed system shall then be determined using the following formula: annual AC output (kWh) = kWp x Kk x SF.

Author's note: The evaluation and assessment approach detailed in the following pages is my interpretation of MCS's standard estimation method. I have not been able to follow MCS instructions, though they are shown here for those who may be able to make more use of them than I am able to. I cannot find out what is meant by 'the appropriate location specific table'* (see right). Also, I have added the variable N% to the formula shown – it is not included by MCS but it seems logical that any variance from '100%' as indicated in the illustration should be included in the annual output calculation.

Site evaluation

The most useful and accurate estimate of the performance of a PV system can only be carried out on-site, and an informed customer can make a major contribution to ensuring the efficiency of a planned PV system –

Assessment 1. Establish the electrical rating of the PV array in kilowatts peak (kWp).

Assessment 2. Determine the array orientation and the array pitch in degrees. Use pic 15 to approximately determine the potential % available to your array (N%). This figure applies to the UK Midlands – average optimal tilt angle will vary according to latitude. Find comparable percentage figures if you are in a different country or in a different part of the UK.

Assessment 3. Look up kWh/m² (Kk) from the appropriate, location-specific table.

Assessment 4. Determine the shading factor of the array (SF) according to any objects blocking the horizon, using the shade factor procedure set out here.

Calculation. The estimated annual electricity generated (AC) in kWh/year of installed system can then be calculated using the formula: annual AC output in kWh = kWp x N% x Kk x SF, taken from the above figures.

Assessment notes

Assessment 1. This will be the planned output for your PV array.

Assessment 2. Array orientation and inclination: orientation (azimuth angle). The general effect of variations in array orientation and inclination on system performance is shown below.

a. The orientation of the array can be measured with an efficient compass or phone app. The required figure is the azimuth angle of the PV modules relative to due south. An array facing due south has an azimuth value of 0°; an array facing either SW or SE has an azimuth value of 45°, and an array facing either east or west has an azimuth value of 90°. The azimuth value is to be rounded to the nearest 5°.

b. The inclination value (ie the angle to the horizontal in degrees, often determined by roof angle) is to be rounded to the nearest 1°C.

Assessment 3. Figures for the UK can be estimated on the map shown on page 80. Much more detailed information for every part of the world is available at some online sites such as www.efficientenergysaving.co.uk/solar-irradiance-calculator.html

Assessment 4. Shade factor (SF). This is a relatively time-consuming business but, unless there are obviously no shading possibilities, it is essential in order to carry out a worthwhile estimate. NB: For systems connected to multiple inverters, or a single inverter with more than one MPP, it is acceptable to carry out a separate calculation of SF for each sub array (ie each array that is connected to a dedicated MPP tracker).

Calculation 1. Where there is an

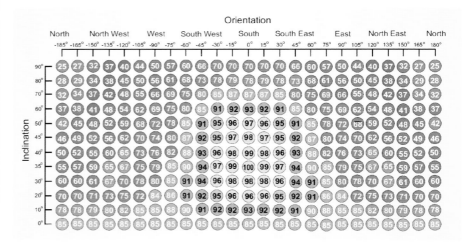

Determine the potential % available to your array. This chart is for a location in central England, and represents the percentage of maximum yield you may expect to get for different angles and orientations. It includes a greater range than the EST chart shown earlier. Incidentally, these and most similar charts do not appear to supply highly accurate information, as the differences between the charts suggest. They should be used for indicative purposes only. (MCS)

obvious clear horizon and no near or far shading, the SF assessment can be omitted and an SF value of 1 used in all relevant calculations.

Calculation 2. This applies only where there's potential shading from objects further away than 10m (32ft) from the centre midpoint of the array. The shading must be analysed and the reading taken from a location that represents the section of the array potentially most affected by shade. For systems with near shading, this will typically be just to the north of the object concerned.

Calculation 3. Where there are objects at or less than 10m (32ft) away from the centre midpoint of the array. It should be used in addition to the method shown for Calculation 2.

How SF calculations are carried out

Important note: This book uses the expression 'shading-edge line' to denote the edge of a shadow area on a chart. MCS uses 'horizon line' in its documentation to mean the same thing. We have selected a different term to avoid confusion between the MCS term and a line denoting the actual horizon.

Calculation 1. You need to assess the potential for shading at all times of the day and year and to bear in mind the potential for tree growth in the foreseeable future.

When there is shading more than about 10m (30ft) away, see picture caption, below –

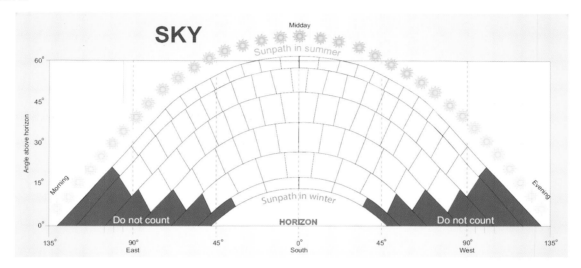

Calculation 2. Shading further away: there are purpose-made instruments for undertaking SunPath assessments, but if you don't have access to these, copy this MCS SunPath diagram. It has a total of 84 segments, each of which has a value of 0.01 (which implies that only 84% of the solar radiation is useable by the PV array). You can now set out to produce a shading analysis, as follows –
1. Stand as near as possible to the base and centre of the proposed array, which might involve standing at an upstairs window, unless there is shading from objects closer than 10m (32ft), such as aerials, chimneys, etc, in which case the assessment of shading must be taken from a position more representative of the centre and base of the potentially affected array position ...

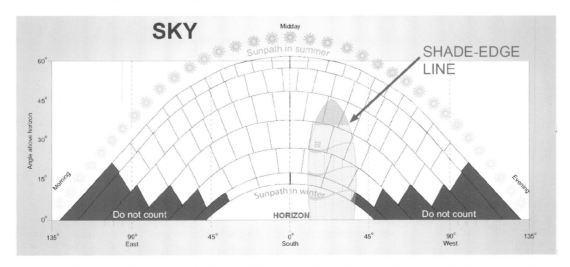

... 2. Looking due south (irrespective of the orientation of the array), draw a line onto the SunPath diagram showing the uppermost edge of any objects that are visible on the horizon (either near or far). (MCS) ...

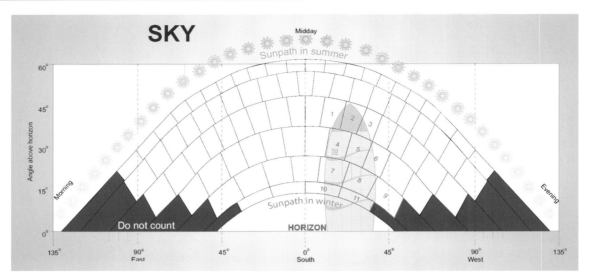

... 3. Once the shading-edge line has been drawn, the number of segments touched by the line are counted. (MCS)
Here, 11 segments are covered or touched by the shading-edge line. The number of segments is multiplied by 0.01 and the total value deducted from 1.0. This provides the shading factor. For example –
$$1.0-(11 \times 0.01) = 1.0-0.11 = 0.89$$

Calculation 3. Shading less than 10m (32ft) away. Where some near shade is completely unavoidable, the following shade analysis procedure must be undertaken in addition to the method described in Calculation 2.
1. A standard shading-edge line must be drawn as

described previously, but now from the actual array location. (If you use ladders or scaffolding, follow appropriate safety procedures!) Now see the next picture caption ...

3. Carry out a similar calculation as that in Calculation 2,

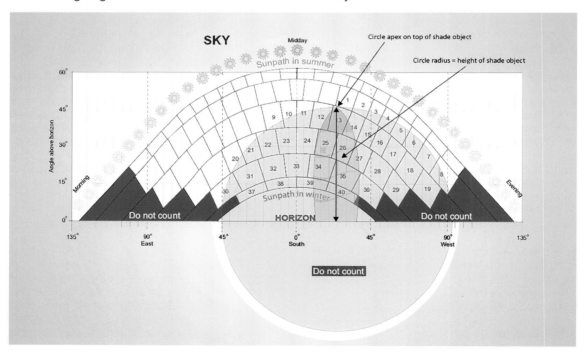

... 2. All objects that are 10m (32ft) or less from any part of the array should have a 'shade-circle' added to the diagram.
• The radius of the circle must be the height of the shaded object from the horizon line.
• All segments touched by or within the shade-circle should be counted as part of the overall shade analysis.
• Where there are several objects within 10m (32ft), draw several circles, one for each object. (MCS).

step 3. Now, there are 40 segments covered or touched by the shading-edge line, so, 40 is multiplied by 0.01 and the total deducted from 1.0, as follows –

1-(40 x 0.01) = 1-0.4 = 0.6

As a result, the object in near shade gives a shade factor of 0.6.

Calculations 2 and 3 combined.

The MCS examples above show – slightly confusingly – the same basic shading-edge lines are shown in both pictures 30 and 31, implying the same or similar shape of obstruction. No mention is made by MCS of what to do if there are objects in both nearer and further away categories, and what follows has been assumed by the author. If it is the case that there are both near and far objects –

- Each one is almost certain to cast its own, separate shading-edge line.
- It seems logical in such cases that the two shading-edge lines are added to the same SunPath diagram and the shaded segments counted – but whenever a shaded segment is affected by both (or more) obstacles, it should not be counted twice.

All quotations and/or estimates to customers will probably be accompanied by one or more of the following disclaimers, where applicable –

- For all quotations and/or estimates: 'The performance of solar PV systems is impossible to predict with certainty due to the variability in the amount of solar radiation (sunlight) from location to location, and from year to year. This estimate – based upon the standard MCS procedure – is given as guidance only, and should not be considered a guarantee of performance.'
- If data has been estimated or taken remotely (without visiting the site): 'This system performance calculation has been undertaken using estimated values for array orientation, inclination or shading. Actual performance may be significantly lower or higher if the characteristics of the installed system vary from the estimated values.'
- Additionally where the shade factor is less than 1: 'This shade

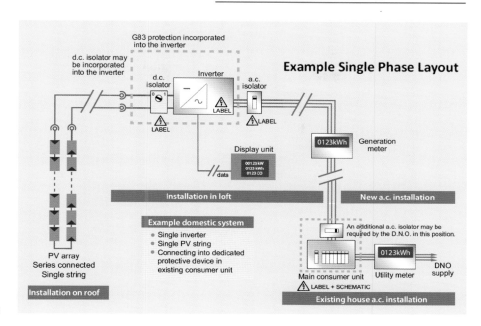

Example Single Phase Layout

G83 protection incorporated into the inverter

d.c. isolator may be incorporated into the inverter

d.c. isolator

Inverter

a.c. isolator

LABEL

LABEL

LABEL

Display unit

Generation meter

Installation in loft

New a.c. installation

Example domestic system
- Single inverter
- Single PV string
- Connecting into dedicated protective device in existing consumer unit

PV array
Series connected
Single string

Installation on roof

An additional a.c. isolator may be required by the D.N.O. in this position.

Main consumer unit
LABEL + SCHEMATIC

Utility meter

DNO supply

Existing house a.c. installation

CONTRACTOR DOCUMENTS. When an installer is employed to carry out a PV project, MCS states that the following information must be supplied to the client before a contract is awarded.
- A performance estimate of the total annual AC energy output.
- The SunPath diagram used by the installer.
- The information used to calculate the performance estimate as illustrated in the table below.

INSTALLATION DATA	Units
Installed capacity of PV system – kWp (stc)	kWp
Orientation of the PV system – degrees from south	degrees
Inclination of system – degrees from horizontal	degrees
Postcode region	

CALCULATIONS	
kWh/kWp(kK) from table	kWh/kWp
Shade factor (SF)	
Estimated annual output (kWp x kK x SF)	kWh

assessment has been undertaken using the standard MCS procedure – it

is estimated that this method will yield results within 10% of the actual annual energy yield for most systems.'

If other estimates or calculations are presented to the client, they must clearly describe and justify the approach taken and factors used, and must not be afforded greater prominence than the standard MCS estimate. In addition, any other estimate must be accompanied by a warning stating that it should be treated with caution if it is significantly greater than the result given by the standard MCS-approved method.

SECTION 2: INSTALLATION
General

The work should always be carried out by an MCS-accredited installer in the UK, or equivalent in other countries. The information in this chapter has been included to reflect the technical standards that MCS registered installation companies are expected to meet in order to gain and/or maintain their MCS certification.

The Renewable Energy Home Handbook

Recommended sequence of working
a. Fit inverter/s in position.
b. Install DC switch disconnector and DC junction box(es).
c. Fit string/array cables – from the DC disconnect/junction box to the inverter, and where each end of the PV string/array will be situated.
d. *Note:* Where it is not possible to pre-install a DC isolator, such as in a new-build project where a PV array is installed before the plant room is completed, cable ends/connectors should be placed temporarily in an isolation box, and suitably labelled as

for an installed DC junction box – see *DC plug and socket connectors.*

When a mains-connected PV installation is being fitted, UK installers must liaise with their Distribution Network Operator (DNO) – the local company licensed to distribute electricity in Great Britain. (See www.energynetworks.org.)

PART 1: INSTALLING THE DC SYSTEM
Most of the work shown here was carried out by leading installer Eco2Solar who have fitted many thousands of PV systems, from the

largest commercial and institutional to the smallest domestic installations. What this company doesn't know about the subject is, quite literally, probably not worth knowing!

1a. Installing inverters
There are many different makes and designs of inverter, and pictured here are two of the best known to demonstrate the general principles of fitting an inverter to a wall. Always follow the manufacturer's instructions, particularly with regard to safety advice.

SAFETY PROCEDURES
An unusually wide range of safety issues apply to PV installations with regard to working with DC current; AC mains current, and working at height, as well as all the usual worksite safety considerations. The main issues associated with PV installation for installers are –

- The supply from PV modules cannot be switched off, so special precautions should be taken to ensure that live parts are either not accessible or cannot be touched during installation, use and maintenance.
- PV modules require a non-standard approach when designing fault protection systems, as fuses are not likely to operate under short-circuit conditions.
- PV systems include DC wiring, with which few electrical installers are familiar.
- PV installation presents a unique combination of hazards, such as electric shock and working at height while simultaneously carrying out manual handling. While roofers may be accustomed to minimising risk of falling or injury due to manual handling problems, they may not be used to dealing with the risk of electric shock. Similarly, electricians would be familiar with electric shock hazards but not with handling large objects at height.
- All persons working on the live DC cabling of a photovoltaic (PV) system must be experienced/trained in working with such systems, and fully acquainted with the voltages

present in that system in particular. Standard health and safety practice and conventional electrical installation practice must apply during the installation of a PV system. Issues such as working aon roofs and standard domestic AC wiring are thoroughly covered by regulatory bodies, and in other publications. There are electrical installation requirements, such as those involving earthing (grounding), islanding (see *Glossary*), insulation, isolation, lLightning protection and others that are specifically laid out by MCS and other regulatory authorities, both in the UK and other countries, and which require the services of a trained, qualified electrician.

Live (DC electrical) working
Live working when installing PV systems is almost unavoidable.
- In every case detailed below, installers should use insulated tools, gloves and insulating matting to stand or sit on.
- Installers can avoid live working by working at night (with appropriate task lighting). Covering an array is not generally recommended because of the problems of access while keeping the array covered during installation, and the effects of wet weather on the covering.
- Working on only one PV panel at a time minimises risk, and is usually made acceptable by the use of appropriate tooling and careful working.
- A temporary warning sign and barrier must be posted for any

period that live PV array cables or other DC cables are being installed.

Electric shock hazard (safe working practices)
It is important to note that, despite taking all of the above precautions, an installer or service engineer may still encounter an electric shock hazard; therefore –
- Always test for voltage before touching any part of the system.
- An electric shock may be experienced from a capacitive discharge – a charge may build up in the PV system due to its distributed capacitance to ground. Certain types of modules and systems, such as amorphous (thin film) modules with metal frames or steel backing, are more at risk. Appropriate and safe live working practices must be followed.
- One example: if an installer sits on an earthed metal roof while wiring a large PV array, he could touch the PV cabling, and may receive an electric shock to earth. The electric shock voltage will increase with the number of series-connected modules.
- The PV array developing a ground (earth) leakage path can cause an electric shock. Good wiring practice, double insulation, and modules of double or reinforced insulation (class II) construction can significantly reduce this possibility, but leakage paths may still occur. Anyone working on a PV system must be aware of this and take the necessary precautions.

92

... and used as a template to mark the necessary fixing positions, which are drilled, plugged in a way that is suitable for the material ...

These SMA inverters (sometimes branded Schüco) were possibly the most popular in Europe at the time of writing. In some cases, such as those with two dissimilar strings, two (or more) inverters will be better than one.

One of Eco2Solar's electricians, Dale, knew he was going to have to mount the inverters on a board – the stone wall was much too uneven. He lined the backing sheet of plywood with non-flammable material, and screwed this to the wall.

These SMA brackets are much lighter affairs. Once again, a spirit level should be used to mark the positions of both inverters, as Dale demonstrates.

... and the inverter mounting panel then screwed into position on the wall.

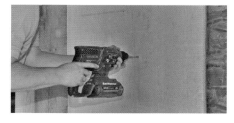

The stone wall was drilled at solid points (not at soft mortar joints), and plugged ...

Retaining screw

On a separate installation, this Fronius inverter has a back plate that is first separated from the rear of the unit ...

All inverters are heavy, and fixings must be very secure. The Fronius inverter is simply hooked onto the backing plate, and fixed in place with its retaining screw.

... and backing plates were attached, as previously.

Again, Dale hooked the inverters onto their backing plate ...

... and secured them with Allen screws. You could use a security fixing, if the inverter is situated somewhere it could be easily stolen.

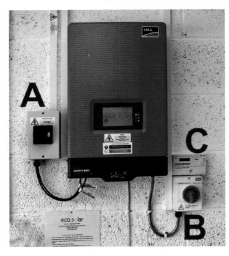

Next, the ancillary items – DC isolator switch(es) (A), AC isolator switch (B), and export meter (C) – are installed in such a way that the wiring runs will be easy and logical to complete. In practice, no two installations will have precisely the same pattern.

Dale used a spirit level to accurately align his wiring clips so that the cables would run vertically and true – an impressive degree of detail.

1b. Installing DC switch disconnector and monitoring

The final ancillary – the AC switch – was installed, but all switches turned off ...

... as Dale removed each inverter front cover ...

... in order to fit the optional SMA bluetooth transmitter to the board inside the inverter ...

... before setting it up to transmit to the required frequency. See Chapter 2 *Insulation, energy saving and monitoring* for information on the SMA monitoring and display system

Here's another (non-Eco2Solar) installation. Because there were two strings of panels, there had to be two DC isolator switches.

Back to Eco2Solar's work again, and these are components installed in a semi-exposed location, manufactured to at least European IP66 specification. This means, IP (or ('Ingress Protection') rated as 'dust-tight' (the first '6'), and protected against heavy seas or powerful jets of water (the second '6').

1c. Installing string/array cables and labels

There are several different types in use, and all must be both IP66 (dust- and weather-proof), and electrically insulated.

After stripping the end of the cable, it was inserted into the correct connector ...

Finally, returning to Dale's work, all of the required notifications and stickers were applied to the installation. At the time of writing, both labels and cables had to conform to MCS's UK requirements –

• Where multiple PV sub-arrays and/or string conductors enter a junction box, they should be grouped or identified in pairs so that positive and negative conductors of each separate circuit may be clearly distinguished from other pairs.

• Great care must be taken to ensure that cable penetrations through the roof and underfelt are weather-tight, and are durably sealed to allow for expected movement and temperature.

• Where long cable runs are necessary (eg over 20m (65ft)), labels should be fixed along the DC cables. Every 5-10m (16-32ft) is considered sufficient to identify the cable on straight runs, where a clear view is possible between labels. The label should state: Solar PV array cable high voltage DC – live during daylight.

It is good practice to label the terminals with coloured insulation tape so there can be no mistakes when connecting up.

... and crimped on with the special crimping tool

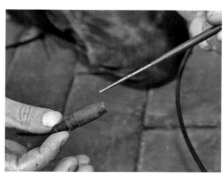

Next, the mandrel was inserted into a connector insulator ...

Purpose-made positive and negative cables ready to plug into the positive and negative connections on the PV panels, when these are fitted.

CONNECTOR TOOLS. Where manufacturers specify special tools to fit together connectors, ONLY this equipment should be used. Failure to use the correct tooling will result in connections that are not mechanically or electrically sound, and can lead to overheating and fires. These are the tools specified by Eco2Solar, specific to these particular connectors, ready for making cable lengths to reach from the inverters to the roof.

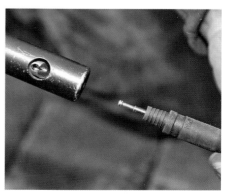

... and pushed into the assembly tool.

The newly-crimped connector was pushed into the tool ...

... which was used to draw together the components.

The end results, for both male and female connectors, will last for many years without breaking down. Because there are several designs of connector, there are several different types of tool for creating connectors – this is just one example.

Justin Walters, working for Eco-Nomical Ltd, used a digital multimeter ...

... to check the voltage output of each panel, just to be 100% sure that each was okay. The point of pre-checking the panels is, if the installer finds a fault after fitting the panels to the roof, there's a lot more work involved in tracing and replacing faulty components than there would be if a fault was discovered beforehand (though, it has to be said, such faults are very unusual).

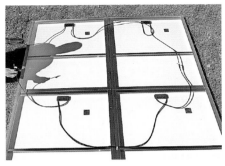

Panels are connected in series, following the circuit diagram supplied with the PV panels and as demonstrated here. There is always a +ve cable (positive) and a -ve cable (negative) on the back of each PV panel. When the cables are connected, the installer must take great care to push the connectors all the way in because loose connections can result in overheating, burning, and other damage.

PART 2: PANEL MOUNTING SYSTEMS

IMPORTANT NOTE: The process of installing panel mounting systems is very similar with solar PV and solar thermal (water heating) systems. Much of the following is common to both, and the relevant material is referred to in Chapter 4: *Solar thermal – domestic water heating*.

This part of the book covers three different types of panel mounting system –

• The most commonly used, on-roof type.
• The so-called in-roof type.
• The rather different free-standing systems, used on flat roofs or for ground mounting.

One common factor that applies to all on-roof systems is the requirement (in the UK) to use and follow MCS-prescribed materials and techniques. It could be that in those parts of the world which receive less rainfall than those in northern Europe, simpler mounting techniques (also noted in the following) may be used – always check with your local regulations.

Note that roof hooks and other fixings designed specifically for use in one country may not be fully appropriate in another. For instance, if a roof hook has been designed for use in a country where thicker roof joists are obligatory, it may not be suitable for use on a roof in other countries

with joists that are narrower. In some circumstances, it may be necessary to add width to the joist around the fixing point, but this must be done in a properly engineered manner with suitably-sized timber, waterproof glue, and rust-resistant screws. The exact sizes and fixings should be specified by a qualified surveyor.

On-roof mounting – fitting roof hooks

The following drawings (unless otherwise indicated) have been reproduced here thanks to Ubbink, the manufacturer of a wide range of approved roof mounting systems.

No roof work should be carried out without scaffolding for obvious safety reasons, but also so that equipment and materials can be handled more easily without causing damage.

The first job will be for the installer to check roof integrity. Quite often, roof repairs are necessary before the process of mounting the solar PV panels can begin. It is also important to inspect the roof underlay for damage during installation works, which should be repaired or the underlay replaced as necessary. Damaged underlay will not provide an effective weather and air barrier, and can affect weather-proofing and wind loads imposed on the roof cladding.

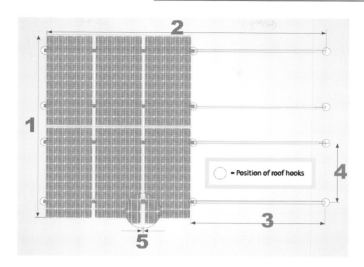

PLANNING THE LAYOUT: The next task involves some research. There are several variables and several fixed points, and these must be calculated in relation to each other. One fixed point is obviously the size of the solar PV panels ...

With a portrait installation, rails should be installed horizontally; with a landscape installation, they should be installed vertically. Ubbink's recommended calculations are –
1. Number of PV panels in the vertical direction x panel height (plus the distance between the vertical panels, if necessary).
2. Number of PV panels in the horizontal direction x (panel width + 20mm) + 50mm.
3. & 4. Horizontal and vertical spacing of the roof hooks – see picture 4. (Also consult the recommendations of the panel manufacturer if other dimensions are advised.)
5. Distance between the PV panels: 20mm (0.8in).

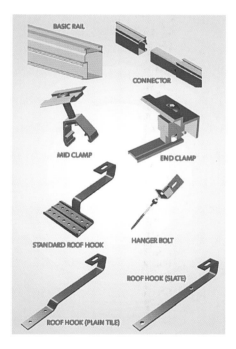

... although their orientation and positioning are not fixed. These are Ubbink's recommendations for rail locations, relative to the PV panels. The Ubbink Solar On-Roof mounting system is a universal-type of panel mounting system, suitable for all slate and tiled roofs, but not for corrugated fibre cement roofs. Different manufacturers have different detail approaches.

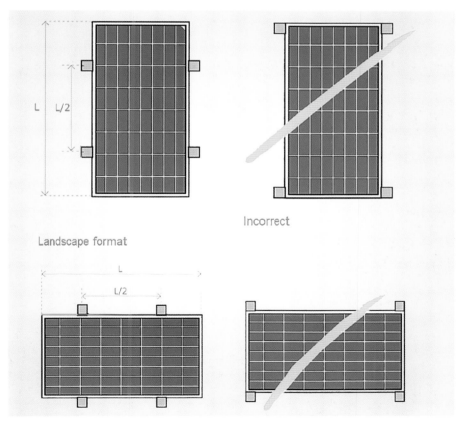

Some manufacturers include a wider range of fitting options than others. This is the Ubbink range, which indicates the types of fittings to choose from.

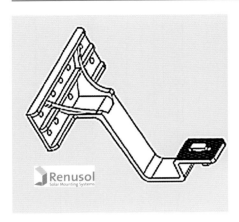

This cast bracket includes a good choice of screw-hole fixings, but it is fixed and will rely on adjustments at the rail fixings to take account of any roof variations.

In the previous shot, Justin was establishing the location of rafters for a through-fixing, which involves drilling the slate or tile. This is no longer permissible under UK MCS regulations because of the likelihood of seals breaking down over time, and because of potential damage to slates or tiles as the roof and solar fittings expand and contract.

As well as removing a few tiles for investigative purposes, you will also need to systematically remove tiles or slates …

… in order to fix the roof hooks to the roof joists. (They must never be fixed to tile battens. Some German manufacturers refer to fixing to 'battens' in their instructions, but one assumes this is a mistranslation of 'joist').

On the other hand, this bracket, used by Eco2Solar on a pantile roof, includes an extremely useful vertical adjustment (arrowed).

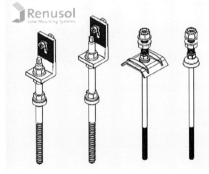

However, for those parts of the world where these things might not matter, through-bolt fixings – invariably in stainless steel – are still available.

Sometimes, there are no roof joists where the roof hook needs to be fitted, and in such cases, the installer will fit noggins between the joists to enable installation of the roof hooks.

Invariably, it is necessary to remove slates or tiles to establish the locations of roof timbers beneath.

Such fixings can be provided with supplementary neoprene seals, purpose-made if required, from suppliers such as RH Nuttall Ltd of Birmingham, England.

The roof hook must never bear on a tile or slate: if necessary, use the packing pieces that were either supplied by or recommended by the hook manufacturer.

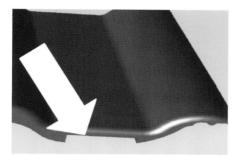

If necessary, use an angle grinder or other cutter to cut a recess in the tile covering the roof hook at the point where the roof hook comes through so that the tile lies flat. If grooved tiles are used, it will also be necessary to cut a recess in the lower tile. This is especially good practice when using thicker tiles such as pantiles because there is otherwise no way for the roof bracket to emerge through the tiles without bearing on one or more of them. If necessary, tiles or slates removed for fixing a mounting bracket should be re-attached with an appropriate mechanical fixing. WARNING: In areas where heavy snowfall is likely, each roof tile that could be touched after deformation of the roof hook under heavy snow load should be replaced with an equivalent metal tile, otherwise the roof tile or tiles might break.

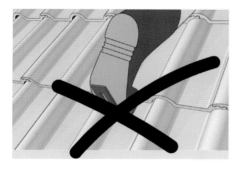

Ubbink makes the point that you should not use fitted roof hooks as a ladder, as this extreme point load could damage the tile below.

With plain roof tiles, a recess must be cut into the tiles around the position of the roof hook.

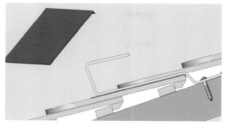

And once again, it's essential that the roof hook does not directly bear on either the tile beneath or above it.

Here you can see where an Eco2Solar installer has removed enough tiles that the roof hook can be screwed to a pattern (or in this case a noggin) beneath (a), and he has also marked the tile (b) ready to be cut …

… and made narrower so that it fits between the roof hook and the next tile to the left.

Now, you can see the space remaining through which the roof hook appears, and the blue line added to the photograph indicates …

… where the last tile to be fitted will finish. The remaining space has been taken up with lead flashing, though see later for the use of purpose-made roof hook flashings, used with roofing slates.

New build

Eco2Solar carries out a large number of solar PV panel fittings to new buildings, particularly on social housing projects such as this.

This is an unusual roof in that it consists of waterproof flat panels, but the principle is the same when working on roof joists. As shown in earlier drawings, you have to establish where the rails – and thus where the hooks – need to be fitted.

My idea was to make a template so that the location of the rail – in this case the mid-point of the slot on the hook – could be marked on the template with a drilled hole (arrowed), while the fixing holes' positions were also drilled into the template.

When the template was placed so that the lower hole was positioned where the mounting rail would be, it was a simple matter to drill pilot holes …

… and install the roof hook by screwing it in place.

Hook types

There seem to be more shapes and types of roof hook than you can shake a tile batten at! Here are three more examples, though many more variations are available.

This is the simple type of hook seen earlier in this section, with only enough offset for slates or thinner, flat tiles.

Where thicker tiles are used, a greater offset will be required. Note that some hooks are specifically produced to cater for roofs with much thicker roof joists, and it's important to ensure that the hooks used are suitable for your roof joists, otherwise joist width may need to be extended.

Used with 'rosemary' tiles, this hook is designed so that the fixing screw passes right through the tile batten, where battens are too close together to allow for the straightforward use of a more conventional bracket. The lower screw must be long enough to penetrate with sufficient strength into the joist – the thickness of the batten doesn't count.

Two more tips –
- Fixing coach screws can be time-consuming and hard work. Make sure you have drilled the correct size of pilot hole, wipe a little grease onto the coach screw thread, and use something like the Makita impact driver (see bottom-left picture) which fully tightens the screws in seconds.
- When using stainless steel roof hooks (the usual material), be sure to only use stainless steel coach screws and washers. Stainless steel actively attacks mild steel, with the result that mild steel screws could easily corrode through, leaving the roof hooks unattached.

Proprietary (solar flash) hook flashings

It is very difficult to create hook flashings that are completely watertight. The not-exactly-modestly-named Genius Roof Solutions has produced a flashing product especially for solar PV and thermal roof hooks, and, at the time of writing, it is the only proprietary product that the author is aware of. This section explains how these flashings are used through the same principles could well apply to similar flashings.

Slate roofs

Working on natural slate roofs usually requires additional skills to those required for other types of roof. In the following, we show the SolarFlash™ roof hook flashing system being used with an existing natural slate roof – the illustrations being screen grabs from the excellent video at geniusroofsolutions.com – and also with the author's new installation of composite polyester 'slates,' which are far easier to work with and have a expected minimum 50-year life span.

The SolarFlash system from Genius Roof Solutions is designed to create a waterproof solution to refitting slates or tiles around solar PV roof mounting brackets. The company claims that the existing method of dressing a flashing material around the bracket is fundamentally flawed, because, usually, the flashing material is in contact with the bracket. This bracket will inevitably move and by dressing lead around the bracket, any movement in the bracket will cause movement in the lead. As the lead is interwoven with the slates or tiles, there is potential for the lead to damage and crack the slates.

So, the key issue is that any flashing must not be in contact with the bracket. The SolarFlash system does not include the bracket, and Genius Roof Solutions recommends that, when selecting roof brackets, the 'elbow' of the bracket for slate roofs provides a clearance of not less than 40mm (1.6in). However, shims can be used for brackets with 30mm (1.1in) clearance, and without this clearance from the rafter, the bracket may come into contact with the roofing material, or might not allow sufficient room for the slates to be refitted under the bracket. (There may be a difference in clearance with Scottish-type roofs, with slates fixed to sark boards, or

where slates are being added to a solid roof surface.)

The SolarFlash™ system includes Genius Roof Solutions' slate fixing hook – the Hallhook™ – which is necessary to secure the final slates in position.

A roofer's tool called a slate ripper is used to rather brutally extract the nails holding existing slates in place. There's no other way: they are hidden.

With two courses of slates removed, the position on the last remaining course of slates has to be marked …

… so that a slot can be cut in the slate …

… allowing room for the hook to be fitted to the joist beneath.

With this new-fit, the roof hook has already been fitted and the (plastic) slates marked …

… and cut …

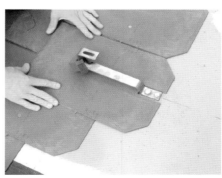

… before being fitted around the roof hook, with an end result similar to that shown in the far left photo.

The Renewable Energy Home Handbook

Natural and cement fibre roofing slates are normally nailed in place through pre-drilled holes. Composite polyester slates can be nailed or screwed without pre-drilling.

Once the next slate in line has been fitted …

… the SolarFlash plate can be placed in position, with the foam seal provided located in the opening that covers the roof hook. Because of the unusual nature of this roof, with no tile or slate battens, (which pushed the roof hook into an unusual position, we had to trim the foam).

The SolarFlash can be simply nailed or screwed in place.

Back to the natural slate roof, where it was necessary to use a plywood packing piece (supplied with the SolarFlash kit) …

… placed between the joist and the roof hook so that there was ample clearance between the underside of the roof hook and the slate beneath it.

After fitting the foam, the SolarFlash plate was installed. Measurement guides in both imperial and metric (see arrows) helped locate it correctly .

The SolarFlash roofer marked the slate to be refitted so that it would fit around the SolarFlash plate …

… after he trimmed it using the traditional method (although an angle grinder with stone cutting disc would be a lot easier to work with for most of us!).

These slates, once trimmed, were simply nailed in position as they were before the SolarFlash was fitted.

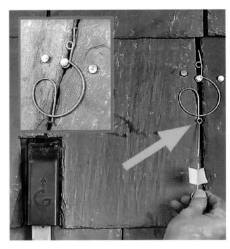

The final slates cannot be nailed in place because of the slates that cover them, and it's not unknown for the slates to be held in place with some sort of adhesive. Sooner or later the adhesive will break down and the slates will slip, leading to a leaking roof. There are various mechanical means of fixing the final slates in place, and a particularly successful and almost imperceptible one when fitted is the HallhookTM, also from Genius Roof Solutions. These are nailed into the gaps between the existing slates, and a temporary puller is hooked on) …

102

... before the final slate layer is pushed into place, taking care not to unhook the temporary puller.

With the slate in position, the puller is pulled downward until the small hook on the bottom of the clip grips the bottom of the slate.

The completed roof hook installation and the refitted slates are now firmly held in position, free of contact with the roofing slates and as weatherproof as before work started.

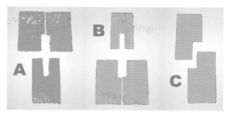

These are SolarFlash's recommended cuts to be made in slates, according to the location of the roof hook –
A Rafter lands where two slates join.
B Rafter lands In the middle of the slate.
C Rafter lands between the middle and the edge of a slate.

An almost identical procedure, except for the ease of working ...

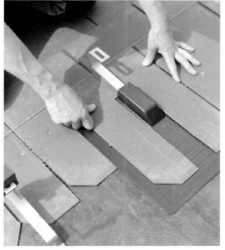

... was carried out on our new roof installation, this being the second layer of slates.

The upper top layer of slates can, of course, be fixed in the conventional way.

This was our completed composite polymer slate roof with SolarFlash flashings around the roof hooks, ready for Eco2Solar to attach the PV panel roof rails

The principle of installing a tiled roof around a roof hook is very similar ...

... to that of the slate roof covered earlier.

Shaped tiles

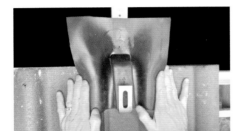

A Solarflex flashing is available for roofs with shaped tiles, demonstrated here on some corrugated roof sheet.

Once again, the principle of installation is very similar to that of the installations shown earlier.

On-roof mounting – fitting roof rails

In an ideal world, it would be possible to adjust roof rails and their mounting hooks both vertically and horizontally. But depending on which make is being used, installation is likely to follow one of the connection systems – or something similar – shown here.

This pantiled roof had been fitted with all of its roof hooks by Eco2Solar, and is ready for the next stage ...

... which is loosely attaching the rails to the roof hooks. In the case of these Ubbink brackets, you must ensure that the hammer-head bolt is vertically positioned in the rail channel after tightening. In principle, the same applies with all types fixing: ensure that the fixings are applied according to manufacturer's instructions.

When there is a slotted hole in the roof hook, this allows optimal adjustment of the height of the rails. This one has horizontal adjustment only.

Position the first rails, aligning them with one another and the roof slates or tiles, and ensuring the rails are parallel. Tighten the nuts used to fasten the rails to the roof hooks/hanger bolts to the manufacturer's recommended torque.

Where longer rails are required, standard rail lengths must be extended with proprietary joining pieces.

Having correctly installed the hammer-head bolts, Ec02Solar's installer, Rob, loosely fitted the locking nuts ...

... and then, when all alignments had been made as described earlier, he tightened them: first by hand and then with a torque wrench.

Showing the mark of an experienced installer, Rob ran the relevant cables along the mounting rails, cable tying them in place. This meant that they were much easier to install than after the panels were in place, and also that they could be tidily held off the roof, which is both neater and helps prevent any build-up of debris under the solar panels.

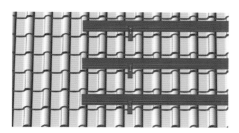

These are Ubbink rails ready for extension.

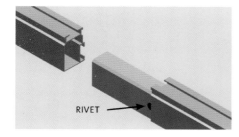

RIVET

On this occasion, the connecting pieces slide into the Ubbink rails, one end before the other, the fitted rivets …

… acting as stops, and ensuring that the connecting pieces slide in by the correct amount.

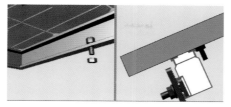

Module installation: Before installing them, Ubbink recommends that the bottom modules are fitted with anti-slip protection (only with horizontal rail installation). To do this, tighten M6 x 20mm screws (with the shaft downward) with M6 nuts in the bottom fastening holes of the module. Lay the modules of the bottom line so that the anti-slip protection of the frame is against the bottom rail.

To install the modules to Ubbink frames, use these clamps, pushing the module end clamp sideways onto the frame rail. The end clamps can be adjusted to the height of the module frame by turning the Allen bolts anti-clockwise using the Allen key, then secured (torque 9-10Nm).

This Eco2Solar installation shows the rails fitted in place, and demonstrates that it is sometimes possible to design the hook locations so that they correspond with the edges of existing tiles. These very thick tiles have each been notched to allow them to sit flat over each roof hook …

… as has this pantile construction roof.

We mentioned earlier that Eco2Solar likes to run DC cabling along the rails as they are installed. Depending on where the cabling has to pass, it may be necessary to create a special roof flashing …

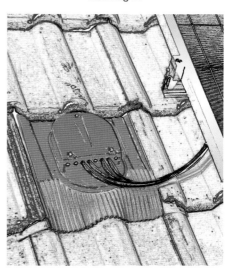

… or your installer may choose to use a purpose-made flashing which provides a watertight lead-through for cables to pass through the roof.

In-roof mounting kits

There are several types of in-roof installation system for both solar PV and solar thermal water heating, and the one thing they all have in common is the ability to integrate solar panels in the roof rather than lay on it, while still retaining the roof's waterproof qualities. The following instructions are for the Ubbink Solar In-Roof system, supplemented by further information for ATAG's solar thermal in-roof system. ATAG's standard in-roof mounting kit is designed to accommodate two or three solar thermal panel systems, and is suitable for most types of roof tile, except slate.

Fitting the panels is an intrinsic part of installing in-roof kits, so this has not been covered separately, as for on-roof panel types.

The Ubbink Solar In-Roof mounting system integrates all sizes and types of PV modules into slate or tiled roofs, ensuring an attractive and low-profile appearance. The system is fully watertight and easy to install, and is typically installed on new roofs, where it reduces cost as tiles or slates are not required below the PV array.

IMPORTANT NOTE: The following are outline details – an overview – of what is involved in installing certain ATAG and Ubbink in-roof kits. Supplied with each kit will be detailed instructions specific to each make and type of kit and relevant to each roof type, which should be followed.

ATAG solar thermal panel in-roof kit

IMPORTANT: Typically, with in-

Typically, an installer will remove the tiles from the roof according to the size of system: eg for a two-panel ATAG installation, a tiled area of 2600 x 2600mm should be removed, and for every additional panel, a further tiled area of 1100 x 2600mm.

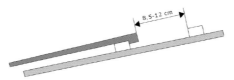

Various support battens have to be attached to the roof as shown in ATAG's detailed instructions.

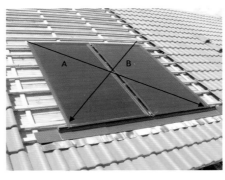

The lower flashings are installed and the solar panels fitted against them.

Panels should be 10cm (4in) from the edge of the bottom flashing.

This is followed by making the connections to the panels; then the remainder of the fixings and the weatherproofing flashings.

When the installer has completed fixing all of the flashing, the roof tiles can be replaced.

For solar thermal systems, the installer will prepare the area where the panels will be mounted, and check where the roof tiles or slates will best fit in.

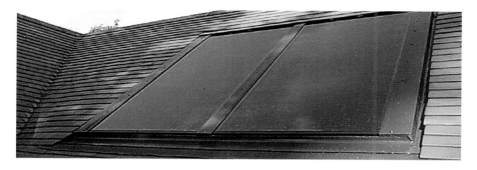

A finished in-roof panel certainly looks neater than on-roof types. With a solar thermal flat panel, such as this ATAG example, there is said to be no drop in performance, though there seems certain to be a reduction in cooling ventilation on the rear of PV panels – and hotter PV panels are less efficient than cooler ones.

roof solar thermal systems, pressure testing must be done before the flashing kit is completed, which must be carried out before fitting the top and side flashings are sited as it may not be physically possible to do so afterward.

Ubbink PV solar panel in-roof kit

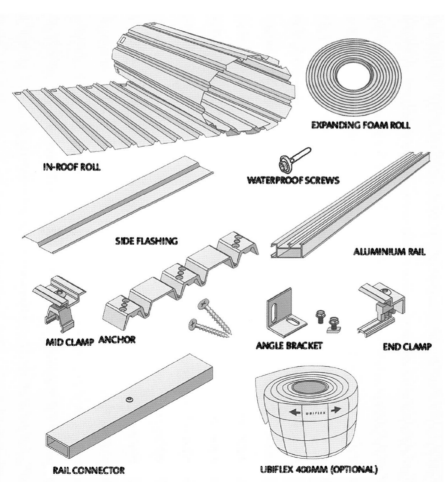

These are the components of the Ubbink in-roof kit.

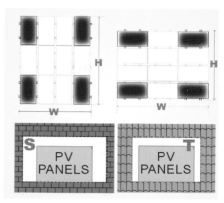

Key: S = slate roof. T = tiled roof. H = height. W = width. PV panels = 'photovoltaic field.'
Roof preparation: This is a similar process as that described for solar thermal panels , except –

• Remove an additional column of tiles on both right- and left-hand sides (two rows for slates).
• Remove an additional row of tiles above the calculated photovoltaic field (two rows for slates).
• Do not remove any additional tiles below the photovoltaic field (for slate leave the first row of hooks).
• Slate roof: remove additional two columns on each side and an additional two rows above the photovoltaic field.
• Tile roof: remove additional column on each side and an additional row above the photovoltaic field.

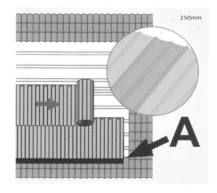

Fit a new batten 240mm (9.5in) above the batten for the row of tiles below the array, then install 400mm wide Ubiflex (A), screwing it to the new batten lapping over the tiles beneath – a waterproof method. Ubiflex is available in black, grey and terracotta.
• Install the in-roof roll by rolling it out (arrowed) and tacking in place.
• Install cable for connection through the overlap, three ridge profiles in.
NOTE: For slates, slip the in-roof roll into the first row of hooks left in place.

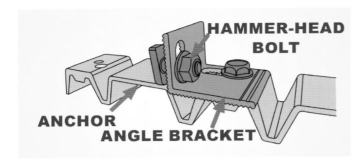

Fix the entire row of anchors (with brackets loosely attached) at the bottom of where the photovoltaic field will go. There are detailed instructions with the kit on where exactly the anchors should be fitted in relation to either portrait or landscape panel formations.

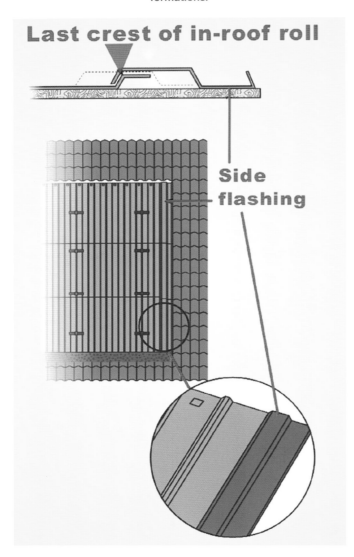

Install the side flashings to make the position between the merge roll and the tiles or slates watertight. Very little of the flashing should be visible once the installation is complete.

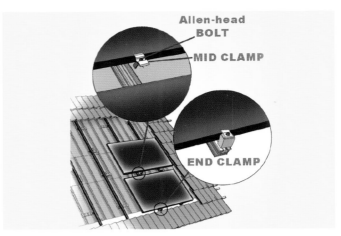

Finally, the rails and panels are installed, following manufacturer's instructions, perfectly aligning the rails so that the panels will be flat, well balanced, and aesthetically pleasing.
• Once the rails are correctly aligned, tighten all nuts and bolts.
• Install PV modules, starting at the bottom.
• Apply the expanding foam roll onto the side flashings on both left- and right-hand sides to the edges of the panels, and to the merge roll just above them.
• When all modules are correctly adjusted and positioned, tighten clamps.
• Replace the rows of tiles or slates previously removed around the side and top of the assembly.

Flat roof and free-standing frames

Where there is insufficient roof space for solar panels, sometimes there is suitable unshaded ground around a dwelling or even a flat roof. All of the following installations have been carried out by Eco2Solar, and demonstrate just some of the possibilities – and limitations.

One limitation is that panels MUST be appropriately attached to the building substructure. They could require weighing down (some authorities quote approximately 1 tonne of ballast per m^2), or be attached at least 600mm (23.5in) into underlying brickwork to prevent wind uplift. A structural engineer can calculate the requirements of your particular location.

This small, adjustable angle, proprietary stand could be suitable for either solar thermal or PV panels, although ...

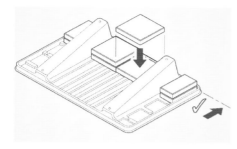

... like the fixed angle Ubbink stand shown here, would need fixing or weighing down with blocks.

... and this timber-framed version show what can be achieved on level and sloping ground respectively.

Ground installations also require that the inverter and relevant switches – to waterproof IP66 standards – have to be situated close to the panels, so that the electricity is converted to AC (alternating current) current before having to 'travel' any distance. Losses on DC (direct current) cable runs are much higher than on those for AC.

PART 3: FITTING PV PANELS
- Remember! DC cables are always live during daylight hours.
- Bypass diodes are installed on the modules fitted to the backs of the panels. They can be damaged by making wrong connections to the cables, bypass diodes, or junction box.

This was photographed during part of an actual PV panel installation by Eco2Solar's impressively thorough fitters, Louis and Rob.

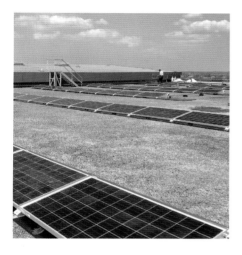

Rooftop mounts, such as this large array installed by Eco2Solar at Wolverhampton University, should be adjusted to provide the optimum angle for the region in which you live.

Even a field which is full of grazing alpacas could be used. These animals have no doubt been specially trained to stand in the shadow, so as not to impede the sun's rays, and to eat unwanted grass from around the panels. Alpaca trainers can be found under 'Cat Herders' in most telephone listings ...

Connect earth and DC cables for PV array

When the first solar module is fitted, the positive cable and the negative connector on the DC extension cable are pushed together.

This ground-mounted, galvanised steel structure ...

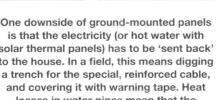

One downside of ground-mounted panels is that the electricity (or hot water with solar thermal panels) has to be 'sent back' to the house. In a field, this means digging a trench for the special, reinforced cable, and covering it with warning tape. Heat losses in water pipes mean that the panels will need to be relatively close to the house, and so technical calculations of pipe size, water flow rates, pipe lengths and insulation would be required.

As the second panel is installed ...

... the negative cable of the first PV panel is connected to the positive cable of the second solar PV panel. Then, from the second solar PV panel, connect the negative cable to the positive cable of the third solar PV panel.

Continue in the same way until all the PV panels for the array have been connected. The negative cable of the last PV panel in the array is then connected to the positive connector on the other DC extension cable; usually running back to the isolator switch next to the inverter.

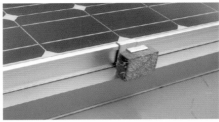

TIP 1: Frequently, depending on the fitting system, panels are temporarily fixed with a double-clip intended to work in conjunction with the next panel. Use temporary packing pieces to ensure the first panel is held firmly while preparing the next.

- *TIP 2: It's usually easy to recognise -ve (female) and +ve (male) terminals, but in case of doubt – or for cables where end terminals have yet to be fitted – it often helps to colour-code the cable ends.*
- *TIP 3: After connecting all solar PV panels, installers are advised to again use a digital multimeter to check the voltage output of the complete array to check for any wiring mistakes, recording the measurement results and comparing them with the manufacturer's test figures.*
- *TIP 4: Cables must be held up to the mounting rails with cable ties so that they don't touch the roof. Otherwise, debris such as leaves can be caught around cables, allowing a build-up that can cause a roof leak.*

It is best if all connections have a durable label (ideally in yellow) stating: 'DANGER! DO NOT DISCONNECT UNDER LOAD' because of the risk of arcing. All cables must be fitted securely and neatly in the roof space, and through to the isolator/inverter. The mounting frame must be properly earthed/grounded (and connected to a lightening arrester if required) in accordance with local electrical installation regulations.

PART 4: CONNECT TO AC SUPPLY

CONNECTION TO MAINS VIA CONSUMER UNIT: This work must only be carried out by a qualified electrician with relevant training and accreditation. This is Eco2Solar's electrician, Dale, making connections to the isolator switch – the generation meter is already in place. Of course, PV supplies require both DC and AC isolators.

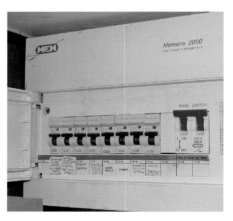

PV supplies should be connected to a dedicated circuit at the consumer unit – ie must not share a final sub circuit. On PV supplies with a residual current device, the RCD must be double-pole.

visible, and C). produced and fixed so that they remain intact and legible for the lifetime of the system. Many labelling requirements have already been shown within the text, and in addition to these, the following labels must be fitted –

MCS also says: 'At the point

Sample Dual Supply Label G83/1

Dual supply labelling should be provided at the service termination, meter position, and all points of isolation between the PV system and supplier terminals to indicate the presence of on-site generation, and the position of the main AC switch disconnector. (MCS)

of interconnection, the following information is to be displayed (typically all displayed on the circuit diagram) –
• Circuit diagram showing the relationship between the inverter equipment and supply.
• A summary of the protection settings incorporated within the equipment.
• A contact telephone number for the supplier/installer/maintainer of the equipment.
• It is also good practice for shutdown and start-up procedures to be detailed on this diagram.'

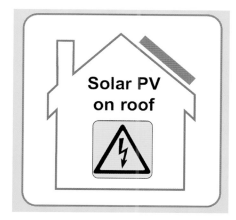

PV SYSTEMS FITTED ON ROOFS ONLY
To ensure that fire and rescue services are aware that a PV system is installed on the roof, this sign must also be fitted next to the supplier's cut-out in the building, at a size of at least 100mm x 100mm (4in x 4in). (MCS)

Ensuring labels stay in place
It is a very common mistake for installers to fit labels without preparing the surfaces to which they are attached, with the result that the labels will peel off in time. Surfaces must be clean of dirt, dust, and the grease that transfers from bare hands. Wipe away obvious dirt, then wipe every labelling surface with silicone remover, methylated spirit, or similar solvent. Use a fresh area of rag and a fresh application of solvent each time.

PART 5: INSPECTION, TESTING, COMMISSIONING AND DOCUMENTATION REQUIREMENTS
This is something that your approved installer will invariably arrange, and usually includes –
• Electrical installation certificate.
• Inspection and testing of AC circuits: documentation for the AC side typically comprises three items.
• Inspection and testing – DC side (including PV array).
• Continuity test of protective earthing (grounding), and/or equipotential bonding conductors (if fitted).
• Polarity test.
• String open circuit voltage test.
• String short-circuit current test.
• Functional tests.
• Insulation resistance of the DC circuits.
• Schedule of items inspected.
• PV array test report.
• Schedule of test results.

Acknowledgement is readily given to the Microgeneration Certification Scheme (MCS) as copyright owner for material used in this chapter. MCS illustrations have the suffix (MCS). The MCS is supported by the UK's Department for Energy and Climate Change, and has produced an industry-led certification scheme for microgeneration products and installation services.

• *AUTHOR'S TOP TIP – Just go for it: If it seems as though all the earlier, high-falutin' calculations for measuring panel angles are just too much, there is no need to get too hung up about it! Provided that the orientation (compass bearing) and tilt angle are within the generous limits for your location, and there isn't a significant amount of shading, it'll probably be fine. (But do take experienced, specialist advice.) Generally speaking, PV is quite forgiving in its demands...*

Chapter 7
Wind turbine installation

This section has been produced with the assistance of Leading Edge Turbines Ltd (LE Turbines), whose knowledge and professionalism has been the most impressive in our preparation for this chapter.

Different installations and circumstances may require that this work be carried out in a different way to that shown here. It is essential that the installer reads and understands all of the installation procedures relating to the wind turbine and tower being constructed before starting any of the work.
(Energy Saving Trust)

We have deliberately omitted details of two types of wind turbine –
• Vertical axis wind turbines. These have blades that spin vertically rather than horizontally like an aircraft propeller. They take up less space and are arguably better looking – certainly more modern – than traditional wind turbines. However, the amount of energy that these extract from the wind is more dependent on rotor size than any other part of the turbine. It doesn't matter which way round it's spinning, if there's less of it, it will gather in less energy. Moreover, it's pretty certain that vertical axis (sometimes called paddle-type) turbines are inherently less efficient than propeller-type ones. You don't see any commercial ships or aeroplanes using paddles

instead of propellers – apart from a handful of tourist paddle steamers from a more inefficient age …

• Turbines mounted on buildings. In a study carried out by the Energy Saving Trust, none of the building-mounted turbines it looked at worked efficiently, and some of them actually consumed power rather than generated it because of the 10W used by their inverters during the lengthy periods when the turbine was not generating any energy. In addition, the amount of vibration given off by a turbine makes it highly likely that long-term damage will be caused to the structure of most buildings to which they are attached.

However, vertical axis wind turbines can have a use on leisurecraft which, by their nature, are often in very windy environments, and where often only a relatively small amount of power is required for battery trickle charging. In fact, in seaborne conditions, there may be a requirement for robustness and compact use of space over efficiency. The LE-v150 is said to be virtually indestructible, even in extremely high winds – but this book is about buildings rather than boats …

PART 1: PLANNING

Wind turbines have a unique place in the renewable energy arena. They are the Marmite or Vegemite of the renewables world: most people either love 'em or hate 'em. For those who love them, it can be difficult to make an objective decision about whether or not a wind turbine will be right for them. The problem is that the general perception of wind, and its measured power over the long-term, are not necessarily the same. LE Turbines says that customers sometimes claim they 'know' they live in a windy spot because the space between their house and next door's feels like a wind tunnel. But just because somewhere often seems windy it doesn't necessarily follow that there will be enough, consistently delivered wind energy to make a wind turbine viable. So, finding out whether or not the installation of a wind turbine will be practicable and cost-effective is the first and most important step to take.

Will a wind turbine work for me?

The Energy Saving Trust has carried out a field trial of wind turbines actually installed in the UK, and says: 'The turbines monitored were domestic, small-scale, building-mounted and free-standing turbines, ranging in rated power output from 400W to 6000W. A turbine's rated power indicates its output (in Watts) at a specified wind speed. Most manufacturers have, to date, chosen to rate their turbines at wind speeds ranging from 11 to 12.5 m/s. Customers should be aware that manufacturers are not yet required by an industry standard to rate their product at an agreed wind speed, and that turbines begin to produce energy at different wind speeds (known as the 'cut-in' speed). In this report, wind speed is expressed in metres per second (m/s).'

The theoretical maximum efficiency of any wind turbine is 59%. This is known as the Betz limit. (In 1919 Albert Betz, a German physicist, calculated that it is not possible for the wind to convert more than 59.3% of its kinetic energy, when turning a wind turbine, into mechanical energy. This is quite separate from individual generator inefficiencies: it's a theoretical maximum.)

The peak efficiencies of the turbines in the Energy Saving Trust's field trial, as claimed by the manufacturers, range from 30 to 60%. The measured peak efficiencies of the turbines monitored in the field trial ranged from 30 to 45%. Results from the field trial illustrate that the building-mounted turbines did not approach the commonly quoted load factors of 10%.

Building mounted turbines: No urban or suburban building-mounted sites generated more than 200kWh per annum, corresponding to load factors of 3% or less.

Free standing turbines: The best performing free-standing sites in the field trial were remote, rural locations, usually individual dwellings situated near the coast or on exposed areas such as moorland.

CHOOSING A SITE

Wind turbines work best in exposed locations, without turbulence caused by trees or buildings.

Planning and other permissions

In the UK alone several different requirements exist, depending which country or region you are based in. In certain areas, some wind turbines are what are known as a Permitted Development. Complications occur when more than one turbine is required, and it is more a certain height from the ground. One of the things that your local planning authority (LPA) might consider necessary – especially if the hub height of the turbine exceeds 15m (49ft), or more than two turbines are proposed – is an environmental impact assessment of the proposed turbine. Even if the height is less, it may, after consultation with stakeholders such as the Environment Agency and Natural England, request that a bat or bird survey be carried out, which can be prohibitively expensive.

The Energy Saving Trust also recommends that: 'Before making an application for planning permission, you should discuss your plans with your neighbours and other third parties who may have an interest, in order to address any concerns that they have. It is also advisable to contact your local planning authority before submitting a planning application to discuss the information it will require with the planning application on other planning issues, such as –
- Visual impact.
- Noise.
- Impact on local heritage (listed buildings and archaeology).
- Ecology (particularly bats).

Measure wind speed

Why is measuring wind speed so important? Wind speeds are difficult to predict and highly variable. The Energy Saving Trust recommends that potential customers first utilise the best available wind speed estimation tools, and then, where appropriate, install an anemometer to measure wind speed across different conditions.

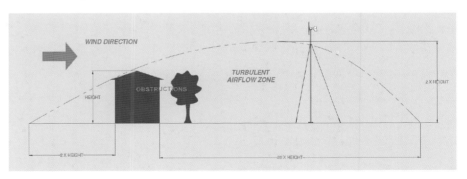

The amount of electricity a wind turbine can generate depends on local wind speed, which, itself, depends on a number of factors, such as those shown in this diagram from LE Turbines –
- **Where you are situated.**
- **Whether there are any obstructions nearby, such as trees and buildings (which slow the wind and cause turbulence).**
- **The height above ground level: wind speeds increase with height, so that the higher a turbine is, the more electricity it is likely to produce.**

Where a wind turbine is sited is, therefore, crucial for maximising overall performance.

Wind power theory

Although the power carried by the wind is proportional to the cube of the wind speed, the actual power output delivered by a wind turbine is more complex. Power output is zero up to the cut in wind speed (the speed at which power is generated), and flat above the 'rated' wind speed. However, between the 'cut in' and 'rated' wind speeds, power output is roughly proportional to the cube of the wind speed.

Wind speed prediction

Online wind speed prediction tools have a reputation for being over-optimistic. What's more they are, of course, incapable of having any knowledge of one of the most important factors: the presence of surrounding trees, buildings, hills or other obstructions. However, they might just be useful for noting whether there is any chance of there being enough wind in your vicinity. For example, the Energy Savings Trust's website has a Wind Speed Prediction Tool, which says: 'Just enter your postcode and say whether your area is rural, suburban or rural, and we'll predict your wind speed and advise on whether a turbine may be suitable for your home.'

The consensus, in our case, was: 'Your predicted wind speed for ABC 123 (rural) is 5.8 metres per second. A domestic small-scale wind turbine may be suitable for your property: however, wind speeds are dependent on local topography and obstructions surrounding a property.' Unfortunately, having tested wind speeds with an anemometer for a winter-inclusive period of around six months, we're pretty sure that a wind turbine would not be cost-effective in our case.

The Energy Saving Trust says 'We do not recommend installing a domestic small-scale wind turbine in areas with wind speeds of less than 5 metres per second (5m/s) as speeds below this are unlikely to provide a cost-effective way of producing electricity with current technologies.

'If the Wind Speed Prediction Tool predicts that the wind speed at the location selected is 5m/s or above, we recommend that, as a next step, you cross-check the result by registering with the Carbon Trust, and using the Wind Yield Estimation Tool at the Carbon Trust website. This tool lets you choose the hub height and the height of the surrounding canopy, and if you select a particular wind turbine and enter power curve data (available from the turbine manufacturer), you should get not only the predicted wind speed but an estimate for electricity generation and carbon savings, too.

'The estimates are based on a simplified model of wind speeds and local physical conditions, which may be more complex in reality, and have a significant impact on the performance of the turbine. So if the project still looks viable, the next step is to check wind speed predictions using an anemometer or wind gauge. You should do this for at least three months, and ideally for twelve months or more. If you measure wind speeds for less than six months, you will need to apply a seasonal adjustment factor as wind speed varies by season.'

If you want to measure wind speed you can –
• Buy an anemometer and a data logger and interpret the data yourself, or
• Instruct a consultant or installer to do this for you – the bigger the potential investment the more likely it will be worthwhile employing an experienced third party to do this work for you.
• If you are already in contact with an MCS-certified installer, this is something that you could discuss with them. MCS-certified installers are required by the Microgeneration Installation Standard to undertake a three-step calculation to assess the likely performance of a small wind turbine system.

Planning permission – the erection of a met mast is not a permitted development in England, Wales or Scotland. Contact your local planning authority before installing a mast to determine whether or not you need to make a formal application for planning permission.

Cost

A reliable way of measuring wind over

WHAT IS AN ANEMOMETER?

The purpose of an anemometer is to measure average, minimum and maximum wind speed, as well as to indicate how much turbulence there is at the site. If two anemometers are placed at different heights on the same mast, this provides useful additional information about wind shear (the difference in wind speed at different heights).

You will also need to measure wind direction. You can do this by using a separate weathervane (also called a wind vane or direction indicator), although some anemometers, such as the Power Predictor and Pro Anemometer, include a direction indicator. Ultrasonic devices can also have built-in wind direction monitors, although these are more expensive.

The pole, anemometer and wind vane equipment are often referred to as a meteorological mast, or met mast for short. Information on wind speed and direction is collected by a data logger, and can be analysed using computer software. The wind data collected also needs to be cross-checked for accuracy against data from a nearby Met Office weather station.

More professional data loggers not only measure wind speeds but also do real-time calculations with that data over regular intervals, usually set at 10 minutes. These calculations include the average and maximum wind speeds over the interval selected.

The ideal scenario is to have an anemometer located at the same site and height as the hub of the proposed wind turbine, so you can leave it in situ while wind speed is monitored.

a period of time will cost you money, expenditure that will either have to be written off if there is not enough wind, or recouped from the electricity generated, if sufficient wind is available. Build this into your calculations.

When comparing manufacturer data, you'll notice that the rated power (in kW) of different wind turbines is given at different wind speeds measured in meters per second (m/s). This makes it difficult to make a side-by-side comparison of likely output from one wind turbine compared with another, even if both state that their rated power is, for example, 5kW. To overcome this problem, RenewableUK has introduced the Small Wind Turbine Performance and Safety Standard, against which the approved wind turbine has to be measured as part of the MCS approval process.

This additional information will help you assess the likely performance of one turbine compared to another –

- The BWEA Reference Power – the rated power of the wind at 11m/s.
- The BWEA Reference Annual Energy – the amount of energy in kWh that the turbine will produce in a year at a constant windspeed of 5m/s at a stipulated hub height.
- The BWEA Reference Sound Levels at 25 and 60m rounded up to the nearest decibel (dB) from the turbine.

CARRYING OUT THE WORK

If you've established that a wind turbine is viable, what an individual homeowner can do himself will be limited by experience and knowledge of groundwork and basic engineering skills. However, when it comes to carrying out electrical work, in the UK and in most other territories, it is illegal for an unqualified person to carry out most aspects of this.

Some installers are happy with the end user to help, especially with regard to preparation and groundwork, and this could be the best way to go for those who want to be as 'hands-on' as possible – provided they have the ability to do so, of course! Areas where this might be possible include –

- Obtaining planning consent (in the UK and perhaps elsewhere).
- Digging the foundation hole and either completing the foundations or working with a contractor to construct the foundations.
- Digging the cable trench.
- If the components are produced for home assembly, fitting and assembly of the turbine head, blades and associated components.

PART 2: GROUNDWORK

Smaller wind turbines can be supported by a guyed tower

In all cases it is essential to ensure that there is suitable access for the delivery lorry and its off-loading equipment, and (in some cases) a crane and/or work platform.

(essentially a pole with guy ropes), like that used for Leading Edge Turbine's LE300 or LE600 units, and even larger wind turbines can be supported by a heavier-duty version of the same thing. Most larger wind turbines, however, are supported by a purpose-built tower for which a substantial concrete foundation is required.

Large-scale towers require much more substantial foundations. The quality of the foundation is of crucial importance to the safety of

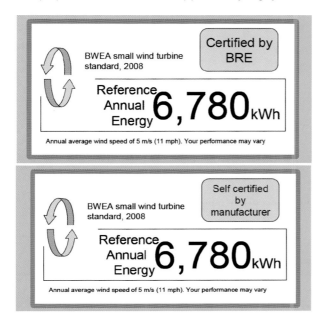

A turbine which meets these requirements should have a label (which should be used on all product literature and advertising) that looks like this.

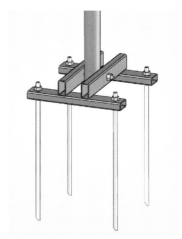

The base support supplied for LE Turbines' lighter-duty turbines consists of a very simple frame, a pivot point for the pole, and rods which are driven or buried in the ground. Different types of rod are available to give varying degrees of grip in different ground substrates.

This is the same base support fixed into a small concrete foundation.

the wind turbine as a whole, and design specifications must be strictly followed. The foundations for a wind turbine tower must be designed to carry the loadings indicated by the turbine manufacturer. These loadings will be passed into the ground and, so that a foundation's shape and dimensions can be properly calculated, a ground condition survey (and, if necessary, a soil test analysis) may have to be carried out – follow your supplier's advice.

There are several different foundation types, but the most common are made from reinforced concrete, cast accurately to enable precise connection to the turbine tower. Of course, not only must the foundations remain in place with no risk of uprooting, but the foundation and its fixings must also be strong enough to support the structure.

There will also need to be a trench in which to run the electrical cable from the turbine tower to the control

Once the required dimensions of the foundation have been established, your contractor can start work. Do bear in mind that you will need to provide adequate access for the equipment he will use, as well as for delivery of ready-mixed concrete. The other point to bear in mind is that soil removed from the ground must be legally disposed of. Once it's broken up into clods and loose soil, it can often take up twice the volume of the hole itself.

box housing, which is usually in an adjacent building.

PART 3: TOWERS
This section looks at the installation of

These contractors are constructing a shuttering frame around the top of the hole. In the UK, there are restrictions on the depth a hole or trench can be without supporting the sides to prevent cave-in while being worked on.

The concrete must be reinforced with purpose-made steel reinforcement positioned according to the correct specifications, and securely tied together so that this won't be displaced as the heavy liquid concrete is poured in.

This particular LE Turbine tower has to be bolted to the ground. The bolts are mounted in a wooden carrier, shown here being fitted to the framework. Note the hole in the middle of the carrier: this is to allow the electrical cable – which will run down the centre of the tower – to be cast in situ.

Obviously, it's preferable to get everything done during daylight hours, but sometimes this is just not possible. Note the use of the underground soil pipe in this installation to carry the electrical cable out of the side of the foundation and into the adjacent trench. Should a cable need to be replaced in future, it will be relatively easy to pass it along a pipe of this size.

Concrete pouring must be carried out to industry specifications to prevent surface and thermal cracking – the only way to ensure that the foundation will be of the required strength and standard.

It is imperative that the formwork is not removed until the concrete has fully set because of the risk of disturbing the bolt fixing points in the concrete.

Here, concreting is complete, the bolt threads are ready to be cleaned up, and the electrical cable awaits arrival of the tower.

This turbine tower is fitted to the ground via a socketed tube let into the ground rather than bolts. The process is essentially similar: the tube has to be cast into the concrete block, and the electrical cable made ready along its centre.

Here, you can see a turbine tower base that has been lowered onto the fixing bolts and is being tightened down.

Whether the tower has a base, as in the previous photograph, or a tube cast into the ground, an earth (ground) connection needs to be made in accordance with manufacturer's specifications.

The inside of the base or socket can also be used to house a junction box connecting the underground cable to the one that will pass inside the turbine tower.

This concrete pad is being disguised with a layer of soil which will be grassed over. Of course, it will need a lot of watering in dry conditions.

both lighter-weight towers supported with guys, and larger, self-supporting towers. It is not intended to supplant manufacturers' instructions, but is an overview of typical work that will need to be carried out.

This section also shows in outline how to install a number of different wind turbine towers supplied by LE Turbines: a procedure that is similar for most comparable towers, although it is essential that the manufacturer's latest instructions are followed when erecting any type or make of wind turbine tower.

Mechanical safety notes
Working with tools of any kind can be dangerous. A tower kit and turbine require some mechanical assembly with basic hand tools, and LE Turbines recommend that this work is carried out by qualified, informed personnel, who should follow the information below –
- Install your turbine and tower kit on a calm day.
- Ensure that the tower kit and turbine are installed in a suitable position where nobody can approach or be

harmed by the turbine rotor blades.

- Ensure that the tower is located in a position where it cannot cause damage to buildings, neighbouring properties or utility lines, should the tower fall for any reason.
- Never attempt to climb the tower: it is not designed to handle this type of loading and may fail.
- If, at any time, a component of the tower or turbine works loose, correct it immediately.
- Always ensure that all personnel in the immediate vicinity are aware of any intended lifting and/or hoisting operations. Check there are no loose components or tools that could fall and cause injury during these operations. Where possible, all assembly work should be completed at ground level.
- Ensure that batteries are disconnected during the installation process.
- Twist together the turbine output cables (to create a short-circuit) during the mechanical installation process. This will prevent the turbine from 'spinning up' during installation and erection of the tower kit.
- When performing routine inspection or maintenance, always stop the turbine using the stop switch.
- Ensure protective gloves are worn when handling guy wires.
- Tower sections can be heavy: ensure protective equipment is worn.

TOWERS WITH GUY ROPES
Some guyed towers are designed for DIY installation, and, in such cases, the manufacturer's information and advice must be precisely followed.

Two people can install the LE Turbines Guyed Tower Kit in a very short time. No winches or gin-poles are necessary, although concrete may be required depending on soil conditions.

The following tools are required to assemble the Guyed Tower Kit –
- 19mm A/F spanner or ratchet (two required).
- Power drill.
- 6.5mm and 13mm diameter twist drill bit, suitable for drilling through steel.
- Long tape measure or steel rule.
- Cement and mixing tools (required for loose or sandy soil conditions).
- Sledgehammer (10kg).

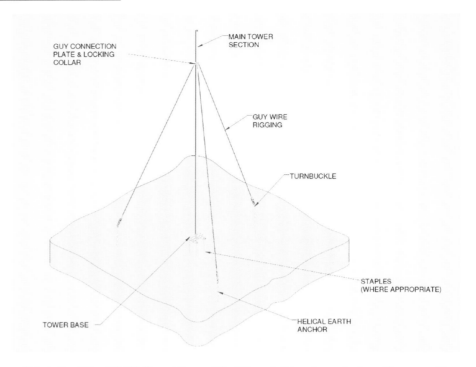

This is the 7.5m (24.5ft) Guyed Tower Kit. Although it has been designed for use with the LE300 Turbine from Leading Edge Turbines Ltd, it is possible to use it with other turbines that require a 48.3mm (1.9in) outer diameter pole (see *Pole selection*), says the manufacturer. However, it does add that under no circumstances should a turbine with a rotor diameter greater than 1.4m (4.5ft) be installed using this tower kit. The moral being: always consult maker's or supplier's specifications.

- Spirit level.
- Gloves.

Pole selection
With this kit, the main length of pole for mounting your turbine is not supplied by LE Turbines as it is not economical to ship. The pole required is a 48.3mm (1.9in) outer diameter x 4mm (0.15in) wall thickness circular hollow section. This is commonly used as scaffolding tube or for fencing applications, and is also known as 1.9in Schedule 20 or 40 pipe. A galvanised pole should be used to prevent corrosion. The tube requires a 6.5mm (0.25in) hole to be drilled in the top end to allow attachment of the LE300, and a 13mm (0.5in) hole to be drilled at the lower end to act as the tower pivot point at the ground mounting point.

Assess soil condition
- Use the table below to decide on the best anchoring system for the tower kit.
- The kit is supplied with four tower base staples and three guy anchors.

- Different fixings may be required depending on soil condition. These are readily available at most hardware stores.
- For permanent installations, it is recommended that all ground fixings are set in concrete. LE Turbines recommends that a purpose-made expansion bolt, such as a suitable version of 'Rawlbolt,' should be used for fixing the tower base to concrete or rock.

See the soil condition table with relevance to tower base fixing and guy anchor fixing (see table on facing page).

Soil condition	Tower base fixing	Guy anchor fixing
Hard solid rock	M12 standard expansion bolt	M12 expansion bolt with eye
Soft solid rock	Long M12 expansion bolt	Long M12 expansion bolt with eye
Loose sand	400mm X 400mm x 200mm deep cast concrete with M12 T bolts or M12 expansion bolt	Helical earth anchor (supplied) fixed into cast concrete
Loose gravel	400mm X 400mm x 200mm deep cast concrete with M12 T bolts or M12 expansion bolt	Helical earth anchor (supplied) fixed into cast concrete
Rocky soil	400mm X 400mm x 200mm deep cast concrete with M12 T bolts or M12 expansion bolt	Helical earth anchor (supplied) fixed into cast concrete
'Grassy' soil/loam	Tower base staple (supplied) direct into earth	Helical earth anchor (supplied) direct into earth

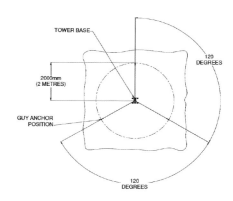

These are the positions of the 7.5 metre tower's anchors. Measure and mark these positions, starting from the position of the tower base. One of the guy anchors must be in-line with the tilting action of the tower to allow easy final erection.

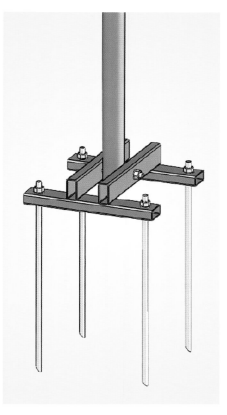

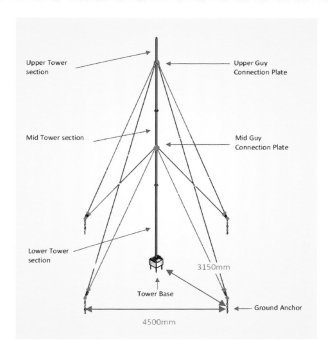

Installation: Mark out the tower anchor positions. This is the heavier-duty, 9m (29.5ft) Guyed Tower for LE300 or some heavier-duty turbines. Begin by selecting an appropriate position for the tower base. Remember that the guy wires will radiate out from the centre of the tower, so you will need to ensure there's enough room: a 4 meter (13ft) diameter is required for the tower guy wires.

Once you have selected the appropriate type of ground anchor, begin by positioning and fixing the tower base. If a cast concrete raft (foundation) is required, dig and cast the raft according to the dimensions shown in the foregoing table. The concrete raft should be made from 80 Newton concrete. Do not shutter or backfill.

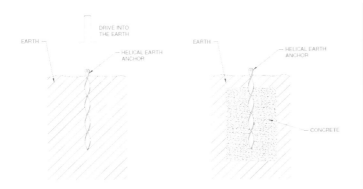

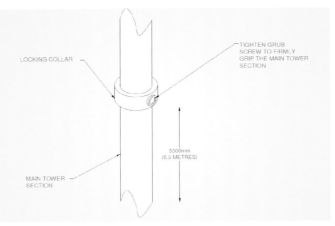

Using the positions marked out earlier, secure the guy anchor bars in the ground, concreting them if required. If the guy anchor is being driven directly into the soil, ensure that the guy anchor bar is driven into the earth vertically, or at an angle pointing away from the tower.

On 7.5m Guyed Tower Kit models, slide the locking collar along the main tower section from the top end so that the top is 5300mm/208in (5.3 meters/17.4ft) from the tower base, and tighten the grub screw with an Allen key to secure it. Slide the connection plate down from the top of the tower (7.5m Guyed Tower Kit models) until it meets the locking collar.

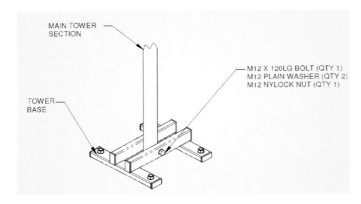

With the appropriate holes drilled in the main tower section, and the tower base firmly fixed in position, insert the end of the tube with the 13mm hole into the base and secure it with the M12 x 120 long bolt. The main tower section can now be pivoted to rest upon the ground to carry out the remaining assembly stages. It is recommended that the top end of the main tower section is lifted slightly above the ground by placing it on the tower kit packaging, as this will make the next assembly stages easier.

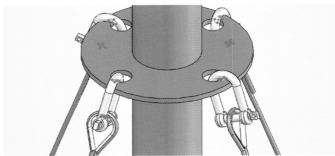

Arrange and fit the guy wires to the guy lugs. With this type of fitting, the crimped end of the guy wire should be attached to the tower using a D-shackle, leaving the uncrimped end to be adjusted to length to suit the anchor position.

Measure the relevant amount of steel catenary wire for each of the guy wires (dependent on model).

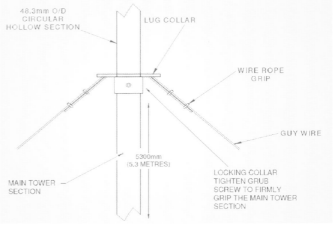

With this lighter weight-type of fitting, allow a 300mm (11.8in) overlap, and always use two wire rope grips at each looped end of the wire. Repeat this for all of the remaining guys to the upper and mid connection plates.

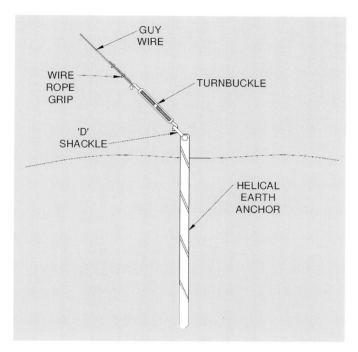

Prepare the fittings on the earth anchor side of the guy wires, allowing a 300mm overlap, and always use two wire rope grips at each looped end of the wire.

Connect two of the guy wires to the guy wire anchors, leaving the single guy wire that works in line with the tilt-up direction. Using the 'D' shackles, attach these two guy wires to a corresponding guy anchor.

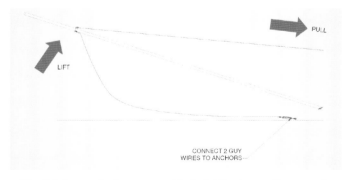

The third guy wire is now used to hoist the tower. Two people may be required for this: one to pull the guy wire, while the other lifts the tower from the other side.

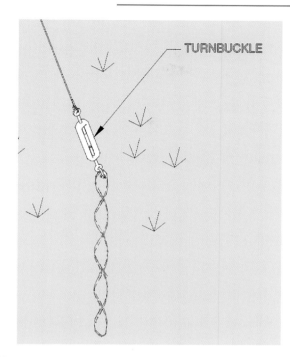

Use a spirit level to ensure that the tower is vertical. Slowly tighten each turnbuckle to achieve a good tension on each guy wire. The tower must be securely fixed in a vertical position with no side-to-side movement. Adjust individual turnbuckles to ensure the main tower section is vertical. If required, the guy connection plate position can be adjusted by moving the locking collar.

Assembly

With larger, multi-section towers, assemble the tower sections. Ensure that any excess build-up of galvanizing zinc is removed from the connection flanges before assembly so that the sections connect squarely.

Lower the tower again in order to install the turbine, reversing the procedure described in picture 13. The turbine can now be installed following the instructions provided with the turbine, after which the same procedure is followed to re-erect the tower complete with turbine, bearing in mind the extra weight involved.

Maintenance

Following this preventative maintenance programme will ensure that the guyed tower operates safely. Always shut down the turbine before attempting to carry out maintenance on any part of the system.

Monthly checks –
• Ensure that all foundations and earth anchors are sound and secure.
• Check that the tower pivot pin is secure and has not worked loose. Adjust if required.
• Ensure that all guy wire-rigging fixings are secure and have not worked loose.
• Check guy wire tension. Adjust if necessary.
• Ensure that the tower is vertical. Adjust if necessary.

STRUCTURAL TOWERS

This type of tower is self-supporting, and relies on a solid, concrete pad foundation having been previously installed, as described in *Part 2 Groundwork*. In most cases, the base for the tower will have been cast into the concrete pad.

In this installation, the experts from LE Turbines, aided and abetted by the customer, have assembled the tower structure from its component parts and are lifting it into position ...

... onto the base that was cast into the concrete earlier, and onto the hydraulic lifting frame which is temporarily attached to the base.

This is a similar exercise: the red-painted lifting frame can be seen attached to the lugs at the hinge point and at the bottom of the base section.

The hydraulic pump is used to slowly raise the tower ...

... until it is fully in position on the base section ...

... after which the tower is bolted in place, following manufacturer's instructions, and the lifting frame – which you can see here has already been detached from the tower – can be unbolted and removed from its temporary pivot points.

PART 4: TURBINE INSTALLATION

Installing the turbine onto the tower usually takes place before the tower has been fully erected (see the section on tower building and erection). However, part of the tower is always installed before the turbine. The following advice is not intended to replace manufacturer's instructions, but is intended as an overview of the work that will be required.

ESSENTIAL WIND TURBINE SAFETY

What follows is an extract – a guide, for general information only – from the safety precautions issued by LE Turbines in relation to its LE600 model. Safety instructions specific to the wind turbine you are working on must always be followed and adhered to.

Please use common sense when installing and operating your turbine!
    Safety must always be your primary concern during the assembly, installation and operation of your turbine. Always be aware of the risks involved with mechanical and electrical installation work. If in doubt about any issue regarding your turbine, please seek further assistance before proceeding. Installation of the LE600 turbine should only be undertaken by suitably competent and qualified personnel.
    Mechanical safety hazards
- Install the turbine on a calm day.
- The main rotor is the most obvious and serious mechanical safety risk. When the turbine is operating at its rated performance, the blades will be very difficult to see due to the speed of rotation. Never approach the turbine while it is operating; always shut down the turbine first by activating the stop switch. Ensure that the turbine is installed in a suitable position where nobody can approach

or interfere with the path of the rotor blades.
- Working with tools of any kind can be dangerous. The LE600 turbine requires some basic mechanical assembly with rudimentary hand tools. If you are in any doubt about how to use these tools correctly, please seek advice from a suitably experienced person.
- Your turbine will inevitably be installed on a tower or other support mount, which may involve working at height. Always ensure that all personnel in the immediate vicinity are aware of any lifting/hoisting operations that will occur. Check that no components or tools are likely to fall and cause injury during the lifting operation. Where possible, all assembly work should be completed at ground level. In the case of roof mount brackets, an appropriately experienced fitter should carry out the installation with the correct equipment for working at height.
- Ensure that the batteries are disconnected during the installation procedure.
- Twist together the turbine output cables (to create a short-circuit) during the mechanical installation process. This will prevent the turbine from 'spinning up' during installation.
- Never install the turbine upside down or in any orientation other than that depicted on the installation instructions.
- When performing routine inspection or maintenance, always stop the turbine by activating the stop switch.

Electrical safety hazards

The following are general pointers only, and do not cover all aspects of installation because electrical work must only be carried out by qualified personnel, who must follow manufacturer's safety instructions appropriate to the wind turbine being installed.
- Battery systems can deliver a seriously large amount of current. A short-circuit can result in hundreds of amps flowing through the battery cables, causing heat build-up and an electrical fire, ultimately. Batteries can explode when shorted. Electricians must always use insulated electrical tools when working on the battery's electrical connections.

- Batteries are very heavy, so do not attempt to move them by yourself. Always use manual handling tools and enlist the help of an assistant.
- Always store lead acid batteries the correct way up. Do not allow the acidic electrolyte to spill or come into contact with any part of you. Always follow manufacturer's safety instructions when handling lead acid batteries.
- Never run the LE600 'off-load' with the output cables unconnected.
- The magnet rotor within the LE600 turbine is constructed using Neodymium magnets (also known as Neodymium Iron Boron – NdFeB). These rare earth magnets are semi-exposed until the turbine is fully assembled. They are extremely powerful magnets and can cause injury if not handled properly. Take care when working with tools made of ferrous materials (such as spanners and screwdrivers) close to an LE600 alternator. The magnetic forces between ferrous materials and the magnet rotor within the alternator are very strong, prompting a sudden snapping action that can cause injury.

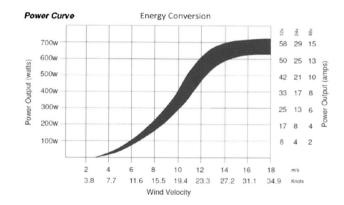

The LE600 is often used as a wind power addition to an existing off-grid power system, to complement Solar PV in order to provide more power in the winter months, and reduce diesel usage. Other applications include remote industrial systems, water pumping systems, and lighting and power in remote farm buildings: any application with a moderate power requirement.
Main advantages are –
- The lightweight turbine head makes it easier to install on high masts in difficult locations, and minimises structural and foundation requirements.
- Low acoustic emissions from the advanced aerofoil blade design.
- Fully 'marinised:' aluminium alloy and stainless steel components, protected from the elements with aerospace-grade coatings and anodising.
- Originally designed for industrial applications, the LE600 has exceptional reliability and durability.
- Flexible electrical system – ideal turbine for future system expansion, which might include solar PV or other forms of renewable energy.

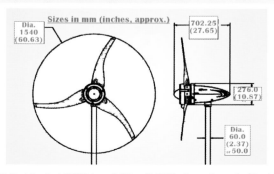

LE Turbine's LE600 is a 1.54m (5.05ft) diameter wind turbine, capable of producing outputs of up to 750W. It has an innovative 'downwind' arrangement which means that the rotor is at the back of the turbine, which gives reduced visual impact, and no tail boom to vibrate and fail through fatigue.

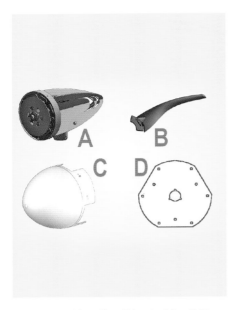

A typical Leading Edge turbine 'kit' consists of the following components –

A   Chassis: 1.
B   Rotor blades: 3.
C   Hub plate: 1.
D   Nose cone: 1.
E   Bag: 1, containing all fixing components (not shown).

This is the very similar-looking component set supplied with a different wind turbine kit.

The following tools are required to assemble the LE600 turbine –

- 10mm A/F spanner and 10mm ratchet (one of each required).
- 13mm A/F spanner and 13mm ratchet (one of each required).
- A set of metric standard hexagon keys.
- Electrical screw drivers.
- Power drill.
- 6.5mm and 3.0mm diameter twist drill bit, suitable for drilling through steel.
- Digital multi-meter capable of measuring DC and AC volts.
- Tape measure or steel rule.
- Thread-locking compound (Loctite 243 or similar).

Connect a digital multi-meter to any two of the three output leads extending from the yaw pivot. With the multi-meter set to detect AC volts (0-20V), a voltage should be displayed when the magnet rotor is spun, which will vary with the speed of rotation. If the magnet rotor rubs, or no voltage is detected while turning the magnet rotor, contact your supplier.

Ensure that the main shaft is free-turning, and does not scrape or rub as it rotates. You may feel a slight resistance from the bearings at this stage. The bearing units used in the magnet rotor assembly are factory lubricated and sealed for life. It will take approximately 100 hours of normal operation for the bearing seals to 'bed-in' and the lubrication to be distributed correctly around the raceways and ball cages. During this period you may notice reduced performance caused by the additional friction of the bearing seals. In operating temperatures of -10 degrees Celsius and below, this 'bedding-in' period will be extended by a further 50 hours of normal operation.

This is the rotor hub plate for an alternative design, seen on another make of turbine.

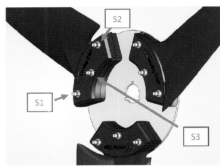

These are the three rotor blades and a rotor hub plate. With all types of turbine, the blades must always be fitted to the same side and to the same orientation. Use the fixing components supplied with the kit. Be careful when handling the blades because they can have sharp edges. Tighten the fixings until the blades are safely secured – and be SURE to follow the instructions specific to the make of turbine installed. Generally, it's important not to over-tighten fixings as this may damage the blades and other components.

With this specific type of turbine (the company that made this one is no longer in existence), the fitter must push the blade sockets into the receptacles in one of the rotor hub sections, being careful to align the pitch pin with the notch in the hub section. In the redundant sockets, between each of the blades, a blanking cap must be fitted.

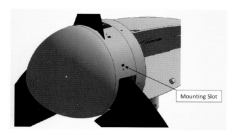

Different types of turbine will have different quantities and types of fixing – always follow manufacturer's instructions.

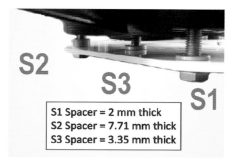

S2 S3 S1

S1 Spacer = 2 mm thick
S2 Spacer = 7.71 mm thick
S3 Spacer = 3.35 mm thick

Some variants of the LE600 require a set of three blade adjustment spacers to be fitted in-between the blade and the hub plate. (See also page 116.) These spacers adjust the pitch of the blade to make it more suitable for the particular variant of the turbine used.

This is LE Turbine's David securing the turbine to the tower. Depending on the type of support structure being utilised for the LE600, it is sometimes advisable to fit the main rotor set to the turbine chassis at this stage. However, in other circumstances, it will be easier to fit the turbine chassis to the support structure before fitting the main rotor set. Depending on the merits of the installation the installer must decide in which order to fit together the turbine, which may mean that the remaining installation steps could follow a different sequence.

Fitting the rotor blade assembly is done by offering the rotor hub plate to the drive hub protruding from the front of the chassis. Three M6 x 16 long cap head screws (arrowed) should be used to fix the rotor hub in position using the three unused tapped holes in the driveshaft. Ensure that all three screws are securely tightened, and that the rotor blades are fitted with the flat side of the blade facing toward the magnet rotor. Thread-locking compound should be used.

The nose cone can now be fitted to the rotor by aligning the mounting slots of the nose cone with the corresponding 'barbs' on the rotor blades. Once the nose cone is fitted correctly over the blades and runs concentrically, use a 3mm drill bit to drill six pilot holes in the plastic block of the blade, using the pre-drilled nose cone holes as a guide. Now, fit six of the No 4 x 12mm self-tapping screws through the nose cone and into the previously drilled pilot holes. (Use an M3 plain washer under the head of each self-tapping screw.) Ensure that the nose cone is secure and true before finally tightening the fixings. NOTE: Where blade adjustment spacers are being used, one of the two 'barbs' will have been removed from each of the blades. Ensure that six self-tapping screws are used to secure the nose cone.

Once again, different turbines will have different arrangements in this regard. Because of the arrangement of this particular Samrey Merlin unit, this is more of a cover guard.

Check the tip spacing. Although the turbine blades are fitted with location 'keyways,' it is important to check tip spacing. Lay the assembled rotor blade on a flat surface, and, using a tape measure or long steel rule, establish that the spacing between each tip is within a tolerance of +/-1.5mm. Adjust the blades as required. Performance may suffer if the blades are inaccurately set.

Check blade rotation. Once the blades have been fitted and secured to the chassis, ensure that they rotate freely. Take this opportunity to check that all of the blade and hub fixings are secure.

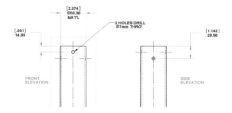

Prepare the turbine mount (see section 4). If you're not using a proprietary structure, it will be necessary to drill appropriate holes in order to secure the turbine. The dimensions shown on the drawing are specific to the LE600, but every manufacturer should provide its own drawings.

Ensure the transmission cables are fitted to the turbine head so that they can be run through the inside of the tower, either before the turbine head has been installed or at the same time.

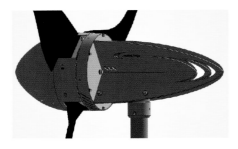

Mount the LE600 turbine onto the support structure. Ensure that the previously-installed power transmission cables are not yet connected to any batteries, and are 'shorted' together (this will prevent the turbine from operating during the installation process). Once this has been done, connect the turbine output cables to the transmission cables using a connecter from Leading Edge Turbines (supplied separately), or a suitable terminal block with a minimum rating of 75 amps (12V), 40 amps (24V), 25 amps (48V), or 20 amps (GT1).

Offer up the turbine to the support structure and push the turbine body onto the tower. Ensure that no cables are snagged. Use the M6 x 75 countersunk set-screws, along with washers and thread-locking compound, to secure the turbine using the holes previously drilled in the support structure.

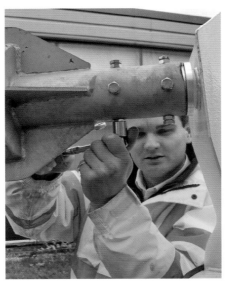

Ensure that the M6 set-screws are securely fastened, tightening them to a recommended torque figure if provided.

PART 5: ELECTRICAL INSTALLATION
Connection to your home

The Energy Saving Trust advises that, in the UK, Part P Building Regulation approval is required for electrical installation and connection of the turbine to your home. Building regulation approval can be obtained by making application to the local authority, or via a member of a Competent Person Scheme (CPS). All MCS installers are required to be a member of this scheme in order to carry out Part P electrical work. Obtaining Building Regulation approval is, therefore, a job for the installer.

Connection to the grid

Your UK installer should liaise with your District Network Operator (DNO) to connect your wind turbine to the local grid. If the wind turbine is up to 16A per phase (equivalent to 3.68kW) it falls under G83/1-1 Stage 1, and your installer can simply inform the DNO within 28 days of commission that a connection has been made.

Larger systems

If your wind turbine system is larger than about 16A per phase (ie it does not fall under G83/1-1 Stage 1), your installer will need to get permission from your DNO before connection to the grid. The DNO will probably carry out a network study (which it may charge you for) to ensure that the local grid network can take the extra power that your wind turbine system will generate. If the local grid network requires work before it can accept your connection, this will be done at your cost. The DNO has 45 days to provide you with a quotation for this work; it must be able to justify the costs it wants to charge (regulated by Ofgem).

You can find key documents about UK grid connection at the Energy Networks Association website (www.energynetworks.org/).

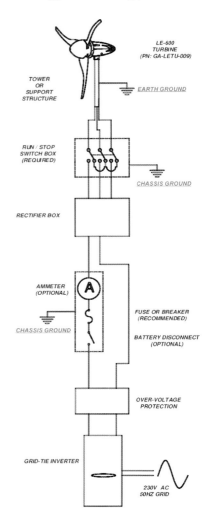

Your electrician should refer to the appropriate wiring diagrams for your installation. This is a typical arrangement (for the LE 600), and the following is provided for general information only, and is not intended to be specific installation instructions for all types of wind turbine.

When the tower or mount bracket is ready to receive its turbine, the next stage is to run the cables from the top of the tower to where the rectifier box/electrical controller and batteries/grid-tie inverter will be located. Your electrician must follow the table below to select correct wire size (cross-sectional area), and must carry out the installation in accordance with local electrical regulations and guidelines.

Careful selection of cable size is necessary: this will vary depending on nominal battery voltage, and the distance that the cables will be run. It will not only affect the safety of the system, but also its overall efficiency. A cable of insufficient size will cause a voltage drop, wasting the power that has been generated. The cable sizes listed below have been selected with efficiency and cost in mind, as it is unlikely that the turbine will be running at full capacity 100% of the time. If in doubt, consult your local electrical supplier.

WARNING: If a cable of insufficient cross-sectional area is used, heat will build up in the cable, constituting a potential fire hazard. Cable capacities quoted are based on 'Tri-rated' cables (BS6231).

Transmission distance

LE600 nominal output Voltage	10 metres	25 metres	50 metres
12 volts	16mm² (3-core)	25mm² (3-core)	50mm² (3-core)
24 volts	4mm² (3-core)	10mm² (3-core)	16mm² (3-core)
GT1 (25-110 volts)	1.5mm² (3-core)	2.5mm² (3-core)	4mm² (3-core)

In a battery-charging system, there may be different ways of wiring together small wind turbines, photovoltaic panels, charge controllers, and batteries. This type of system will often grow 'organically,' but the following guidelines should be followed –

- Wiring of the LE600 turbine and associated electrical systems must be carried out in accordance with national and local electrical codes and regulations.
- Ensure that the turbine is not running or connected to the batteries during the installation or wiring processes. Connect the output cables of the turbine to prevent the rotor from starting.
- Galvanic corrosion of electrical joints. Try to avoid connections between dissimilar metals. For example, connecting copper and aluminium will result in galvanic corrosion of the connection, increasing the electrical resistance of the connection (wasting energy), and reducing mechanical integrity of the joint. Where possible, use a fluxed solder to make electrical joints. Was off excessive flux after soldering.
- The power transmission cables must be protected from mechanical damage and fatigue. Run the cables through an approved conduit/trunking.
- Prevent mechanical strain on the transmission cables running down the tower from the turbine. Clip the cables to the inside of the tower (failure to do this will result in excessive mechanical strain on the cable joints within the slip-ring assembly, and may cause failure). Cable ties or cable glands are a good way to prevent mechanical strain on the cables.
- 'Earth' (ground) the system: the turbine tower should have its own separate earth (ground) point. The negative terminal of the battery bank should also be earthed, as this provides protection against the build-up of static, and lightning strikes. The tower should be earthed separately with its own ground rod if there is a long transmission distance between tower and batteries. An appropriate surge arrester could also be used to help prevent damage to the battery charging system during a lightning strike. Ensure that the earth cables are of the same rating as the positive and negative cables.
- Fuses: the turbine and charging circuit should be protected with a suitably-rated 'slow-blow' DC fuse or DC circuit breaker (refer to the table below for the correct rating). The fuse or breaker should be positioned between the turbine and batteries (on the positive cable). If a stop switch is used (recommended), the fuse should be positioned between the switch and the batteries.
- 'Hybrid' systems – The LE600 turbine can be used in parallel with

LE600 nominal Output voltage	DC fuse/DC circuit breaker rating
12V	60 amp
24V	30 amp
48V	20 amp
GT1	16 amp

Run/stop switch – a simple switch arrangement can provide a safe and easy way of stopping the turbine during high winds or for maintenance. As the switch is thrown, the batteries are disconnected and the turbine is 'shorted,' reducing rotor speed to slow.

Diversion charger (dump load) controllers operate (usually on off-grid systems) by increasingly switching output to a dump load once the batteries begin to reach high voltages. The dump load consumes the 'excess' power from the turbine, which means that the turbine's power output is always utilised, whether or not the batteries are fully charged. Larger capacity battery banks will be able to store more energy, and so less dump load will be used. Do not use a photovoltaic-type charge controller with a wind turbine. For hybrid or more complex systems, LE Turbines recommends that a 45A or 60A diversion controller is utilised, together with a separate dump load. With a grid-connected system, 'surplus' power is returned to the grid.

OFF-GRID: HOW DIVERSION CHARGE CONTROLLERS AND DUMP LOADS WORK

At Leading Edge Turbines, one of the most frequently-asked technical questions is how diversion controllers and dump loads work. After all, it seems a little counterintuitive to have to 'dump' energy in a renewable energy system!

Q. Why 'off-grid'?
A. In most grid-tie systems, unwanted energy is returned to the grid – earning feed-in payments, when available. In some grid-tie systems, however, the dump load is used to protect the grid-tie inverter from damage due to excess voltage. This is most applicable to off-grid systems because dump loads (or diversion loads) are only used to protect the deep cycle batteries from becoming damaged through overcharging. The dump loads aren't usually needed to control or protect the turbine itself. You (or your supplier) must ensure that you choose the correct dump load capacity (sometimes by combining smaller dump loads) for the battery bank voltage being charged. Imagine a typical off-grid system with a small wind turbine charging a bank of batteries. The small wind generator will be continuously charging the batteries according to wind fluctuations. The system will probably also have a number of 'loads' – appliances that consume electricity from the batteries at the same time as they are being charged. The dump load operates with a diversion charge controller, and is much like the overflow in a bath – the bath is the battery bank and the wind turbine is a tap. The tap (wind turbine) continuously fills the bath with water (charge) until the level of the water reaches the level of the overflow (dump load). When this happens, excess water flows into the overflow and away, preventing the bath from overflowing, and your batteries from becoming overcharged and damaged. The dump load facility only comes into action when the batteries are full, so if other electrical loads are draining charge from the batteries, the dump load may activate only infrequently.

Alternatively, if there is very little in the way of electrical loads on the system, and the batteries will be full more often, the dump load will be working more frequently.

All of LE Turbines' diversion charge controllers use a proportional control system. This means that the excess power that is sent to the dump load may be 10W, 300W, 1000W – or something in-between, depending on the size of the dump load. The system quickly and automatically decides how much power needs to be dumped at any one time.

Q. So why not just disconnect the turbine when the batteries are full?
A. Disconnecting the turbine from the batteries and leaving it disconnected is a very bad state of affairs for a small wind turbine, as it will mean that the turbine is open circuit and off-load, and will free-wheel to incredibly high revolutions per minute. This will be noisy – and will mostly likely result in destruction of the turbine!

Q. So why not just switch off the turbine when the batteries are full?
A. Although some other micro turbine manufacturers use a charge control system that switches off the turbine when the batteries are full, there are a number of disadvantages with this. Firstly, it is not especially good for the turbine. Small wind turbines are usually 'stopped' electronically by disconnecting them from the batteries, and then shorting out the turbine (often called dynamic braking). Under high wind conditions, when the batteries are more likely to be fully charged by the small wind generator, dynamic braking can be overcome by the force of the wind, meaning the turbine sends the generated current through the short-circuit and back into itself. This is bad for the turbine, and will often result in a burnt-out stator. Furthermore, frequent stopping and starting of the turbine can be bad for the batteries themselves as it will cycle the state of charge more quickly, potentially reducing the length and quality of service that deep cycle batteries can offer.

Q. Where should the dump load and controller be located?
A. They should be sited as close as possible to your battery charge controller, mounted in free-circulating air but never in an enclosed space.

PV panels. Leading Edge recommends that the PV panels are wired independently with a separate charge controller specifically designed for use with PV panels, and connected in parallel with the battery bank. For hybrid or more complex systems, LE Turbines recommends that a 45A or 60A diversion controller is utilised, together with a separate dump load.

Use of grid-tie Inverters. It is easily possible to connect the wind turbine to a grid-tie (grid-connected) inverter. It is recommended that only grid-tie inverters supplied or recommended by the manufacturer (in this case, Leading Edge Turbines) are used to ensure that an appropriate MPPT curve has been programmed (maximum power point tracking is a technique that inverters use to get the maximum power from a given generator, such as solar panels or a wind turbine).

You may notice the following behaviour during normal operation with the LE600 turbine. General principles apply to other models and other manufacturers, though figures shown are specific to the individual turbine – see relevant manufacturer's data –

- Cut-in – The turbine will not begin to charge the batteries until the rotor is spinning at approximately 290rpm. When operating below this speed, the turbine will be 'off-load' and freewheeling. Once turbine output voltage becomes equal to the nominal battery voltage (at around 290rpm), the turbine will come 'on-load,' and begin to deliver current to the batteries. During the off-load stages of rotation, the blades will rotate very freely. This allows the rotor to build speed and aerodynamic lift to be generated by the blades.
- Normal operation – Once the rotor is spinning at more than 290rpm, current will be delivered to the batteries. As rotor speed increases, so, too, will the current and voltage. Excessive wind speed may increase battery voltage to a high level, and, once this happens, the diversion charge controller will recognise that the battery voltage is too high, and begin 'dumping' power to the heater module.
- Charge regulation – Once the charge controller has switched over to dump load, the turbine will no longer be charging the batteries. Instead, power from the turbine will be delivered to the dump load (usually a resistive heater element). Battery voltage will begin to drop to normal levels during the regulation period, and, once back within acceptable limits, the charge controller will switch turbine output back to the batteries: an arrangement that is more effective than having the charge controller constantly stopping and starting the turbine.
- Shut down – By activating the stop switch, output from the cables of the turbine are 'shorted' together, effectively putting an infinite load on the generator, causing the turbine to stall. When the stop switch is activated, the turbine may still rotate slowly during high winds, but the rotor blades will not be able to build any significant speed. It is not recommended that the stop switch is activated while the rotor is spinning at high speed, as this sudden braking action will stress the blades and other components. Only activate

the stop switch during a 'lull' when the rotor is not spinning excessively fast.
- Very high winds – Due to the configuration of the low-friction, high-efficiency alternator, the electromagnetic braking effect is not as strong as with more conventionally-designed turbines. In certain high wind conditions, the rotor can overcome the electromagnetic braking, which allows high currents to be produced in the stator coils. If this situation occurs with the wind driving the braked alternator for prolonged periods, damage to the turbine will occur. Therefore, the braking switch should only be used to slow the unit prior to manually/mechanically tethering the turbine during very high winds.
- High winds – Every effort has been made to ensure that the LE600 will withstand the forces exerted by strong winds. However, the raw power in high winds is huge, and the stresses placed upon the turbine are magnified by gusty and turbulent conditions. Where possible the turbine should be shut down and tethered in advance of particularly strong winds (60mph+) and storm conditions, as this will minimise machine wear and tear, and will help prevent failure. Protect the turbine from extreme winds as you would other aspects of your property.
- Grid-tie applications – When connected to a grid-tie inverter, the LE600 will operate in much the same manner as when it is charging batteries, except for the use of dump loads because unused power is fed back into the grid. It is important that grid-tie inverter characteristics are matched to the power curve of the turbine to ensure optimum performance. An appropriate electrical interface may also be required, depending on inverter equipment used. It is generally only recommended that turbine manufacturer-approved and programmed inverters are used.

Note: Never allow the turbine to run off-load with no connection to a battery bank or grid-tie inverter, as doing so will allow open circuit voltages to be generated by the turbine, which can be dangerous, and

may damage the stator coils within the turbine.

Maintenance

SAFETY FIRST: Always shut down the turbine before attempting to carry out maintenance.

The following are the maintenance instructions issued with the LE600 wind turbine. Different models and manufacturers will probably have different requirements, so this is an overview of what the owner can expect. Refer to the instructions issued with the turbine you have installed before carrying out maintenance.

Post-installation checks (to be carried out one month after installation). The installer should –
- Check that the tower mount pins are secure and have not worked loose. Adjust if required.
- Ensure that the rotor hub is still securely fitted.
- Ensure that the rotor blades rotate freely.
- Monitor the output. Ensure that the turbine and charge controller are functioning correctly.

Annual maintenance –
• Inspect the tower/support structure.
• Remove the turbine from its installation to a suitable workbench.
• Remove the rotor blade assembly.
• Inspect the edges of the rotor blades for damage such as dents or chips: the blades will become unbalanced if they are damaged, causing vibration, noise and poor performance. If there are many dents along the edges of the blades, a new set of rotor blades should be fitted.
• Inspect the roots of the blades at their mounting positions and the hub mountings themselves for signs of stress cracking or fatigue. New components, such as a new set of rotor blades, should be fitted if any cracks or fractures are apparent.
• Remove any build-up of dirt and debris from the rotor blades using a mild detergent and warm water.
• Check the blade hub fixings for tightness.
• Carefully remove the yaw mount bracket by unscrewing the four connection screws from the sides of the LE600 chassis. Inspect the slip rings and wipers for obvious signs of wear, and replace the wipers if necessary. Carefully reassemble the yaw pivot (do not pinch any cables).
• Check that all electrical connections are sound and free from corrosion.
• Ensure that the turbine is generally in good working condition, and safe for continued use.

Other considerations
• The equipment used in the charging system (batteries, charge controller, PV panels, inverters, etc) should be maintained according to the instructions provided by the manufacturer.
• Where lead acid batteries are used, it is especially important they are carefully maintained, as failure to do so will result in the batteries being rendered useless within a short period of time.

LE Turbines recommends that the rotor bearings and rotor blades be replaced after five years of continuous operation, to ensure that turbine performance and safety are not compromised.

Chapter 8
Ground and air source heat pumps

PART 1: INTRODUCTION

A heat pump works by moving heat, either outside to inside or vice versa. Quite simply, fridges and freezers are heat pumps that transport heat from inside to outside the fridge or freezer compartment. Since domestic heat pumps operate the other way aound – moving heat from the outside to the inside – they are little more than reverse-action refrigerators! And similarly, it is perfectly possible to reverse heat pump technology in order to cool a house instead of heat it.

Because heat pump systems are only efficient when they are generating hot water at lower temperatures than fossil fuel-based central heating systems, a large surface area is needed for the radiators. This can be achieved with underfloor heating or larger-than-usual radiators, Sometimes, fan-assisted radiators are used to increase the rate of heat transfer.

Benefits of heat pumps compared to fossil fuel-powered boilers –
- No combustion process, so no outdoor pollution.
- No need for a fuel store or attention to the stove.

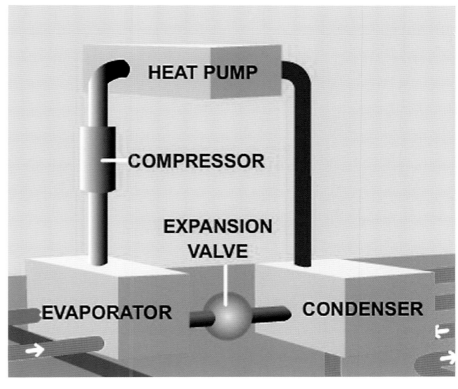

Heat pump technolgy is simple in principle, but can be fiendishly complex in practice. For that reason, we are glad to have the expertise of Mitsubishi Electric UK and Ice Energy to draw on in this chapter, as well as information from the Energy Saving Trust, whose illustration this is.

- No CO$_2$ or other toxic gases inside the building.
- No need for flues or ventilation.
- Long life expectancy.
- Minimal or no annual maintenance.

PART 2: PLANNING

In order to understand how heat pumps work, you may need to take what you think you know about heat and temperature, and throw it right out of your triple-glazed window! It's easy to get hung up on the idea that, beneath 'freezing point' there is no heat. But what we think of as freezing point just happens to be the freezing point of water on the surface of the Earth, which we say is 0°C, or 32°F.

Incidentally, why the difference? Well, the Fahrenheit scale was established by scientists more interested in the freezing point of alcohol (0°F) because of its extensive used in laboratories. Meanwhile, the Celsius brigade rightly decided that the freezing point of water (0°C) has more relevance to our everyday needs.

However, there is another measure of zero degrees and that's what's known as Absolute Zero, or 0° Kelvin, lower than which it's believed to be impossible to go. At that theoretical temperature (which is approximately -273°C), all atomic activity ceases. So any temperature above Absolute Zero has, in theory at least, some heat to give up. Indeed, at 0°C, there is still approximately 273° 'available'! But how to get at it? That's where a heat pump comes in …

Heat pumps aren't magic – they need electricity to run – but what's good is that they extract about three* times as much energy as they consume in electricity.
*The exact coefficient of performance (CoP) varies quite a bit, according to air/fluid temperature and ground conductivity (in the case of a ground-source heat pump). It also depends on how well the system was designed and installed in the first place. A study in 2010 by the Energy Saving Trust, found that most of the 83 devices it monitored for a year were under-performing. About 87% didn't achieve a CoP of 3 which the trust considers the level of a 'well-performing' system (higher is better). And 80% failed to meet 2.6, the level necessary under the EU's Renewable Energy Directive for classification as a renewable source of energy. However, a further study, three years later, showed that design and installation improvements can make a very significant difference. Twenty of the 83 poorly-performing devices received major or moderate alterations, and 17 of them had much better CoP figures as a result. The clear lesson is that heat pumps must be designed and installed correctly, and, if you're spending a large sum of money having one fitted, you want a guarantee from the supplier that it will meet its CoP target range.

Ice Energy is convinced that a heat pump with supplementary domestic hot water electric heating will still be more efficient than an oil-fired, fossil-fuelled boiler. However we have eradicated one potential weak spot from our system by installing different domestic hot water heating systems.

If the system is well-designed in the first place, there's a better than fighting chance it will work well. And I'm pleased to say that our ground source heat pump system, described here, works very satisfactorily, all year round.

Domestic hot water – potential weak spot of a heat pump

In general, it seems that the largest inefficiencies derive from using a heat pump to heat domestic hot water. Heat pumps work most efficiently when producing water at temperatures lower than those required for domestic hot water. Not only is the domestic water of a lower temperature than most people are used to, there is a risk that lower temperature water can contain legionella bacteria. For that reason, domestic hot water systems heated by a heat pump must have a supplementary heat source capable of generating a temperature of at

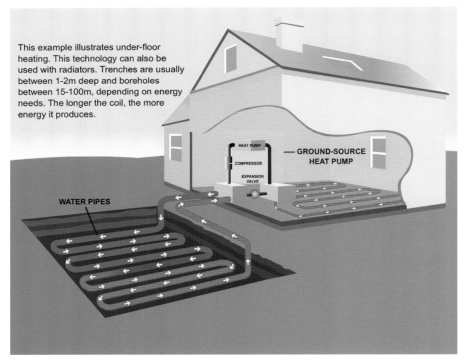

This example illustrates under-floor heating. This technology can also be used with radiators. Trenches are usually between 1-2m deep and boreholes between 15-100m, depending on energy needs. The longer the coil, the more energy it produces.

WATER PIPES

HEAT PUMP / COMPRESSOR / EXPANSION VALVE — GROUND-SOURCE HEAT PUMP

By far the most commonly used sources of heat for a heat pump are the ground, air and, less popularly, a body of water such as a lake. To understand how a heat pump works, you have to remember that, when gases and liquids are compressed, they produce heat. A heat pump transfers heat from the ground, or air from outside, passes it through a heat exchanger, compresses it (in a reverse 'refrigeration' cycle), and turns the basic level of heat in the fluid or air into more heat, which is used to heat the house. The air or fluid being returned is colder than when it came in, having had heat extracted from it. The 'return' flow is sent back to the heat source (the air outside or the pipes in the ground) where it dissipates its 'cold.' (Energy Saving Trust)

least 60°C. Most have a direct electric heater, similar to an immersion heater, which heats water in a cassette within the heat pump, and transfers it to a hot water storage tank next to the pump, or directly to the hot water cylinder. In the UK, this system is obliged to run a 'pasteurisation' cycle to 65°C for one hour once a week. In practice, most of these supplementary heaters work more often, either to top-up the heat in the hot water cylinder to a level preferred by the user, or because the designers have over-specified the timing or duration of the pasteurisation cycle.

In summary –

- With ground-source heat pumps, glycol is pumped round a pipe buried in the ground.
- With water-sourced heat pumps, the pipe lies in a river, lake or (conceivably) the sea, or water is temporarily pumped out of one those sources and then returned.
- For an air-source heat pump, heat is extracted from the outside air.

A huge amount of information can be had from the following UK websites –
1. www.energysavingtrust.org.uk.
2. The 'Department of Energy and Climate Change' website – which has a lengthy address, so best to search for the words shown here in single quotes.

And from the USA, further information is available from –
- www.energy.gov
- www.eia.gov

PART 3: AIR SOURCE HEAT PUMPS (ASHPS)
How does an air source heat pump (ASHP) work?
There are two types of ASHP –

Air-to-air ASHP systems, which have indoor and outdoor units for heating a single room with blown air. These are not covered here as they are not, by themselves, capable of 'home' heating. They can offer air conditioning from the same unit in hot weather, however.

And air-to-water ASHP systems, which transfer the heat extracted from the air outside to water, for underfloor heating and radiators. An air source heat pump can be a highly efficient way to generate your own heat and hot water using stored energy from the air. For more information see www.energyshare.com.

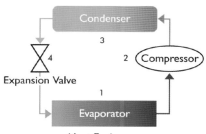

Heat Exchanger

An ASHP works in a similar way to a fridge, except that a fridge extracts heat from inside, and the ASHP extracts heat from outside – even at outdoor temperatures as low as -15°C.

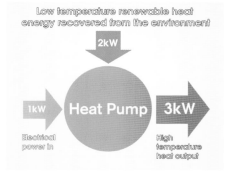

The main benefit of an ASHP system is that it typically produces between 2 and 3kWs of useful heat from every kW of electricity used to operate it. If the property is off the gas grid, an air source heat pump will typically generate lower CO_2 emissions, and facilitate a reduction in energy bills in comparison to any existing oil, electric, LPG, or coal heating system. (Mitsubishi Electric)

Technical details
There are at least three – and usually four – main parts to an ASHP –
1. The evaporator coil, which absorbs heat from the outside air.
2. The compressor, which pumps refrigerant through the heat pump and compresses the gaseous refrigerant to raise it to the temperature needed for the heat distribution circuit.
3. The heat exchanger, which transfers heat from the refrigerant to the water in an air-to-water system.

4. The tank, installed by Ice Energy, is used to store the heated water which, when required, is pumped through radiators or underfloor heating. The heated water created by an ASHP can be used for space heating, and/or for domestic hot water. For all ASHPS you will also need a compatible hot water cylinder for the domestic hot water supply (Ice Energy).

ASHP systems circulate the water round the radiators or underfloor heating at a temperature of between 30 and 45°C, though the lower the temperature, the more cost-efficient the ASHP will be. This temperature offers a more gradual type of heat than a gas boiler, which usually circulates water round the system at anything up to 82°C. These lower temperatures make it vitally important that the underfloor or radiator system is specially designed to heat the house properly at such temperatures (essentially, there just has to be a greater surface area), and that household heat losses are kept to a minimum, with good levels of loft and/or wall insulation.

There are two main designs of air-to-water ASHPs –

- External (monobloc) units: These are made up of one or more one-piece external units – though there is usually only one for domestic use. In mono-bloc systems, all of the elements are packaged into the single outdoor unit while, inside, there is a control panel. This type of system is quite straightforward for an approved ASHP engineer to install. Monobloc units minimise the amount of indoor space required, but the unit does need to be adjacent to the house, and is usually fitted on an outside wall or flat roof.
- Split systems: These consist of an external unit, usually with similar dimensions to an air-conditioning unit (W: 950mm; D: 350mm; H: 750mm approx), and an internal unit about the size of a conventional gas boiler. This type of connection allows great flexibility in the installation of ASHPs, with distances of over fifty metres between the two units possible. Split systems must be installed by an engineer with full F-Gas qualifications (to do with managing fluorinated gases and ozone-depleting substances) because of the refrigerant piping between the outdoor unit and the internal unit.

Planning permission
In the UK, air source heat pump installations require planning permission with regard to their noise level and visual impact of the external unit. Implementation varies with local authority.

Noise
All air source heat pumps contain a fan for circulating air. Some are noisier than others, and they can run both day and night. Check out the noise rating (shown in decibels dB or dBA, at a given distance), and satisfy yourself that this won't annoy you or a neighbour.

Legionnaire's disease
Homes with heat pumps are theoretically vulnerable to Legionnaire's disease but only if the domestic hot water (DHW) is stored for periods below 65°C – the bacteria are killed above this temperature. To prevent Legionnaire's disease, a dedicated immersion heater raises the DHW temperature to 65°C at pre-programmed, timed intervals. However, this will increase your running costs and it's important to get the timings and heat levels right!

AIR SOURCE HEAT PUMP INSTALLATION
The following section has been produced with the kind assistance of Mitsubishi Electric. It is important that only a competent person (such as defined in UK Building Regulations) installs Mitsubishi Electric's heating system (invariably the same stipulation with all manufacturers and suppliers): it is also a requirement of grant and tariff payment bodies. All of the illustrations shown here are copyright Mitsubishi Electric, unless otherwise stated.

Choosing a location for an outdoor unit
The main points to discuss with your installer are –

- Avoiding locations where the unit is exposed to direct sunlight or other heat sources.
- Selecting a location where noise emitted by the unit does not disturb you or your neighbours.
- Selecting a location that allows easy wiring and pipe access to the power source.
- The fact that water may be produced by condensation during operation, and could drip or run beneath the unit. (Occasionally, harmless water vapour produced by the defrost operation may make it appear as if smoke is rising from the outdoor unit. Water vapour normally disappears quickly whereas smoke may not.)
- Selecting a level location that can bear the weight and vibration of the unit.
- Avoiding locations where the unit can become covered with snow. In areas where heavy snowfall is anticipated, special precautions must be taken to prevent the snow blocking the air intake, such as installing the unit at a higher position or installing a hood on the air intake, though note that this can reduce airflow, and the unit may not operate properly as a consequence.
- When installing the outdoor unit, minimum clearances and dimensions are always specified by the manufacturer.

WARNINGS
- Mitsubishi Electric makes it quite clear that its equipment (and this applies to almost all manufacturers) must only be installed by an approved installer or authorised technician, for important technical and safety reasons.
- Failure to comply with approved installation procedures is almost certain to compromise the manufacturer's warranty.
- Units must be installed according to the instructions, and must be securely installed on a structure that can sustain their weight.
- If the internal element of a split air-to-water heat pump is installed in an enclosed area, the qualified installer must take measures to prevent refrigerant concentration in the room in the event of leakage. This does not apply to monobloc systems, where the refrigerant is contained inside the sealed outdoor unit.
- All electric work must be performed by a qualified technician according to local regulations and the instructions given in the manufacturer's manual. Units must be powered by dedicated power lines, and the correct voltage and circuit breakers must be used. Power lines with insufficient capacity or incorrect electrical work may result in electric shock or fire.
- The user should never attempt to repair the unit or transfer it to another location. If the unit is installed improperly, it may cause water leakage, electric shock, or fire. If the air-to-water heat pump needs to be repaired or moved, ask an installer or an authorized technician.

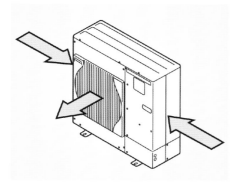

When installing the outdoor unit in a location where it may be exposed to strong wind, the air outlet of the unit must not face directly into the wind. Strong wind entering the air outlet may impede normal airflow, which may result in reduced performance or a malfunction. Where severely strong winds may be encountered, the installer could –
1. Face the air outlet toward the nearest available wall, keeping a distance specified by the manufacturer, or
2. Install an optional air outlet shroud, if available, or
3. Position the unit so that outlet air will normally vent at right-angles to prevailing wind direction.

The fitting engineer will install the unit on a strong, level surface to prevent rattling noises when running. Here, the base for the air source heat pump is being prepared and checked: specially-designed support feet were used.

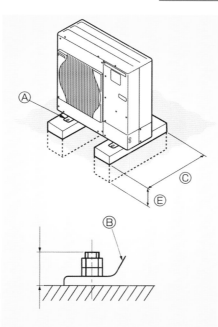

The fitter ensured that the length of the foundation bolt was within 30mm (1.18in) of the bottom surface of the base, and will firmly secure the base of the unit with four M10 foundation bolts. If additional support is needed, the engineer can use the extra installation holes on the back of the unit.
A. M10 (⅜in) bolt
B. Base
C. As long as practicable
E. Set deep in the ground

The heat pump was fitted onto the base, with two installers taking care to check the correct position and ensure that the unit was located correctly.

The distance from the wall was checked for compliance with the recommended minimum space requirement. It is important not to block the vent as operation efficiency will be compromised and the unit may break down.

This will be the final location of this water cylinder, in what will become an airing cupboard on the first floor of the property. A suitably-sized cylinder of the correct specification must be specified when installing any heat pump.

Both external and internal pipe locations were measured ...

... and matching holes were drilled through the wall ...

... for the pipe connections between the water cylinder and the outdoor unit.

The external pipework was connected to the outdoor unit ...

... and the internal plumbing connections were completed as all the internal pipework was connected to the water cylinder.

The initial electrical fitting was completed adjacent to the water cylinder ...

... and this, in turn, has to be connected to the building's electric circuit.

You can see that this control unit is mounted on the cylinder. This illustrates why there is a need for good service access to all components, bearing in mind that ...

... after completion, the finished cylinder would be contained inside a newly-constructed airing cupboard.

As noted earlier, it is essential that all piping is lagged appropriately.

The external water pipes were connected ...

Before beginning the test run, the installer will –
- Turn on the main power switch more than 12 hours before beginning operation. (Beginning operation immediately after turning on the power switch can severely damage the internal parts.) The main power switch must be kept turned on during the operating period.
- Not touch the (uninsulated) refrigerant pipes with bare hands while the unit is running (split systems only). The refrigerant pipes can be hot or cold, depending on the condition of the flowing refrigerant, with the attendant risk of a burn or frostbite.
- After stopping the equipment, wait at least five minutes before turning off mains power, otherwise it may cause water leakage or breakdown.

The base of the outdoor unit must be checked periodically to ensure it is not loose, cracked or damaged. If such defects are left untreated, the unit could fall down and cause damage or injury.

Installing a split system Hydrobox

Only competent refrigeration engineers who have attended and passed the requisite F-Gas and manufacturer product training, and have appropriate qualifications for installation of an unvented hot water Hydrobox specific to their country, should install or service this unit (other manufacturers stipulate the same). All of the illustrations shown here are copyright Mitsubishi Electric unless otherwise stated.

... and appropriately lagged, and the outdoor unit was ready for connection to the electricity supply. Installing the system is something that must only be carried out by trained, accredited engineers.

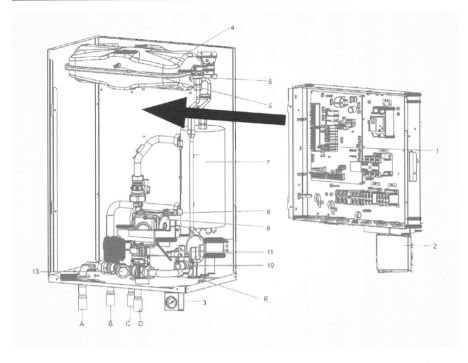

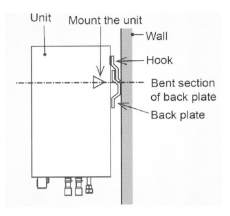

After screwing the backplate to the wall, the installer will lift the unit, align it correctly, and fix it to the backplate.

These, for reference. are the internal components of a Mitsubishi Electric Ecodan Hydrobox, in this case the EHSC series, which provides a good idea of the degree of engineering that goes into making a top-class air source heat pump. None of these components is user-serviceable.

Number	Component
1	Control and electrical box
2	Main controller
3	Manometer (pressure gauge)
4	Expansion vessel (if required – depending on model)
5	Expansion vessel charge valve
6	Automatic air vent
7	Booster heater
8	Drain cock
9	Water circulation pump
10	Pressure relief valve
11	Flow switch
12	Plate heat exchanger
13	Strainer valve
A	Inlet from space heating/Indirect DHW tank (primary return)
B	Outlet to space heating/Indirect DHW tank (primary flow)
C	Refrigerant (liquid)
D	Refrigerant (gas)
E	Discharge from pressure relief valve (installer to connect to suitable drain point)

Hydrobox location

This is another area that can be discussed with your installer, although location will be restricted by the following technical points. The Hydrobox should be –
- installed indoors in a frost-free, weather-proof location; not exposed to water/ excessive moisture.
- Positioned on a level wall capable of supporting its filled weight (shown in the manual).
- Positioned with minimum distances around and in front of the unit to allow service access.
- Secured to prevent it being accidentally knocked over or dislodged.

Primary water circuit

The installer will –
- Thoroughly flush pipework of debris, solder, etc, using a suitable chemical cleansing agent, before connecting outdoor unit.
- Flush the system with water to remove the chemical cleanser.
- Add combined inhibitor and anti-freeze solution to prevent damage to pipework and system components.

Pipework

Following the manufacturer's instructions, the installer will start by checking, before installation –
- Pressure relief valve.
- Expansion vessel pre-charge (gas charge pressure).
- Some heat pumps require the installation of an hydraulic filter or strainer (field supply) at the water intake.

Connection to most versions of the Ecodan Hydrobox should be made using 28mm compression fittings, although the installer will always follow manufacturer instructions. For instance, the Ecodan ERSC series has G1 (male) thread connections. Cold and hot water pipework will not be run close together where possible, to avoid unwanted heat transfer.

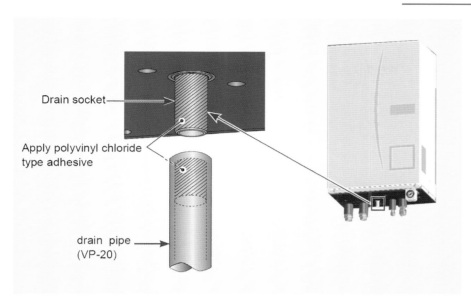

Drain socket

Apply polyvinyl chloride type adhesive

drain pipe (VP-20)

When fitted, a drain pipe will be installed (depending on heat pump model), at a down slope of 1% or more, to drain condensing water. To install the drain pipe, the engineer will apply polyvinyl chloride-type adhesive over the shaded surfaces inside of the drain pipe and on the exterior of the drain socket, then insert the drain socket deeply into the drain pipe, using a pipe support to prevent the drain pipe falling from its socket.

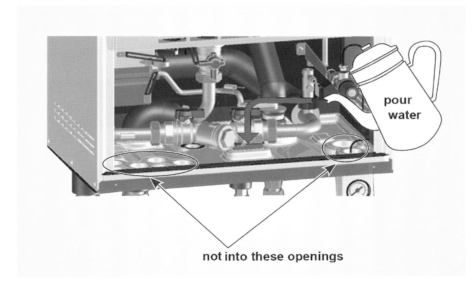

pour water

not into these openings

After installation, the installer will check that the drain pipe drains properly by slowly pouring water (so that water does not overflow) into the drain pan.

Safety device connections

The Ecodan Hydrobox contains a pressure relief valve, from which must be connected appropriate discharge pipework (in accordance with local and national regulations). Failure to do this will result in a discharge from the pressure relief valve into the Hydrobox, causing the product serious damage.

Pipework insulation

- All exposed water pipework, including the pipework and connections at the top of the Hydrobox, should be insulated to preclude risk of burning, unnecessary heat loss, and condensation.

Anti-freeze solutions

These must be used and MUST consist of the correct grade of propylene glycol. Ethylene glycol is toxic and must NOT be used in the primary water circuit in case of cross-contamination of the potable (drinking) water circuit.

Filling

The installer will –
1. Check that all connections (including factory-fitted ones) are tight.
2. Insulate pipework between the Hydrobox and outdoor unit with a special type of waterproof pipe insulation material, with a thermal conductivity of ≤0.04 W/mK.
3. Thoroughly clean and flush the system once more.
4. Fill the Hydrobox with potable (drinking) water.
5. Fill the primary heating circuit with water plus anti-freeze and inhibitor, using a filling loop with double-check valve to avoid back-flow contamination of the water supply.
6. Check for leakages; fix if necessary.
7. Pressurise system to 1 bar.
8. Release all trapped air using air vents during and following heating period.
9. Top up with water as necessary. (If pressure is below 1 bar.)

Refrigerant pipework

This is connected by a qualified F-Gas engineer, in accordance with the outdoor unit's installation manual. In some cases, the refrigerant pipe between the outdoor unit and the Hydrobox has to be connected with a pipe-size adapter.

Electrical connection

All electrical work must be carried out by a suitably qualified technician (a legal requirement in the UK and many other countries). There are two 'immersion-type' heaters. The on-board booster heater and the house immersion heater (in the HW cylinder) should be connected to dedicated power supplies, independently to one another.

Dip switch functions

All ASHP units have switchable options. On the Ecodan, some are controlled by dip switches: small, white switches located on the printed circuit board. There are four sets which must be set by the installation engineer, although there are preferences which should be discussed with the homeowner.

Controller options

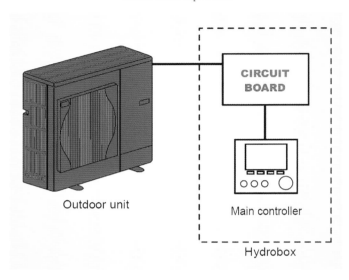

Outdoor unit

Main controller

Hydrobox

The Ecodan Hydrobox, like almost all ASHP internal units, comes factory-fitted with a main controller. This is for temperature monitoring, and it has a user interface to enable setup, viewing of current status, and input scheduling functions. The main controller is also used for servicing purposes (accessed via password-protected service menus). The main controller can be mounted remotely from the Hydrobox.

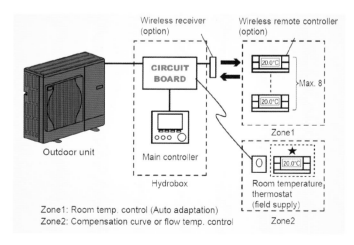

Zone1: Room temp. control (Auto adaptation)
Zone2: Compensation curve or flow temp. control

However, for best efficiency Mitsubishi Electric recommends using a room temperature function, based on a room thermistor fitted in a main living area. This can be done in a number of ways, with several options for both end-user and installer.

SD memory card

Usefully, the Ecodan Hydrobox is equipped with an SD memory card interface which can simplify main controller settings as well as store operating logs.

After installation

Within the first couple of months of installation, remove and clean the Hydrobox strainer, plus any externally-fitted ones. This is especially important when installing on an existing system, where 'aged' debris is almost certain to be disturbed.

Annual maintenance

The indoor Hydrobox must be serviced once a year by a Mitsubishi Electric-trained technician with relevant qualifications and experience. Any electrical work should be done by a tradesperson with the appropriate electrical qualifications. Any maintenance or DIY fixes carried out by a non-accredited person could invalidate the warranty, and/or result in damage to the Hydrobox, and injury to a person.

PART 4: GROUND SOURCE HEAT PUMPS
Checklist

The following is a list of factors to consider when deciding whether or not to have a ground source heat pump installed –

1. Have you got sufficient land? As a very crude rule of thumb, a house with four or five bedrooms might need a hectare (a couple of acres) from straight pipe runs, half of that required for 'slinky' (coiled) pipes. However, a bore hole (much more expensive) requires relatively little land area.
2. Can you put up with the disruption? Unless you have an adjacent field, your entire garden could be devastated for several months, until grass grows back.
3. Is your soil the right type? (See page 141.)
4. Is your house very well insulated – up to the latest new-build insulation standards?
5. How will you heat your domestic hot water, if not via the heat pump?
6. Has the system you prefer got controls that are easy – or even possible! – to understand.
7. If you haven't already got underfloor 'wet' heating, are you prepared to take out all of your existing central heating radiators, and have new, much larger ones fitted?

You may decide that there are too many disadvantages to having a ground source heat pump, and it's certainly not a task to be undertaken lightly – as my wife and I know, since it's mainly our own system that you'll see being installed later on in the book. However, we are very pleased with the system we have in place.

SOIL EFFICIENCY

Here's what to consider when deciding whether or not the soil beneath you is suitable for a ground source heat pump system.

Our ground is 'saturated clay' throughout the winter months, and I'm sure that has a major affect on the efficiency of the system. If the ground is not able to conduct heat easily, the pipes there will not be able to extract heat efficiently from the surrounding area.

In these cases, longer pipe runs in the ground will be

required in order to compensate for the reduced potential for heat extraction. Whether or not it makes sense to do this can only be determined by an experienced professional, and the final calculation will have a bearing on whether a ground source system is cost-effective and practicable for you.

You may consider seeking professional advice on the thermal conductivity of your soil. However, accurate determination of ground thermal conductivity is not easy – there are so many variables. You would need to consider the degree of compaction, moisture content, and the presence of moving groundwater, all of which have massive influence, and all of which are subject to variation. Some suggest that the best you could ever hope to achieve is a good approximation. With that in mind, here are some useful pieces of information.

Ice Energy points out that limestone and chalk, frequently found in some parts of the UK, are not included in these charts but have very low thermal conductivity.

Typical thermal conductivity of different soil types (the higher the number, the better!)

Soil type	Watt/metre Kelvin*
Dry sand or fine gravel	0.76
Loam	0.9
Clay	1.1
Silt	1.66
Saturated clay	1.66
Saturated sand	2.49

*This is the standard (SI) unit of conductivity.

How to identify soil texture

Remember to check the soil at the depth you are planning to install the GSHP pipes.

Soil Texture type	General characteristics				
	Main constituents	Appearance; dry conditions	Rubbed between fingers: dry conditions	Squeezed in palm of hand: very moist/wet conditions	Rolling into ribbon: moist conditions
Sandy soil	At least 85% sand particles	Crumbly with no clods or lumps; individual soil grains visible to the naked eye	Gritty: soil grains readily felt	Can form cast; crumbles with least amount of handling	Cannot form ribbon
Sandy loam soil	At least 50% sand particles; not more than 20% clay	Mainly crumbly and loose; grains readily seen and felt	Gritty: soil grains readily felt	Can form cast that will bear careful handling	Cannot form ribbon
Loam soil	At least 80% sand silt in about equal proportion; not more than 20% clay	Mainly crumbly; some clods or lumps	Fairly smooth but some gritty feeling; lumps easily broken	Can form cast that can be handled freely	Cannot form ribbon
Silt loam soil	At least 50% silt; not more than 20% clay	Quite 'cloddy' but some crumbly materials	Lumps easily broken and easily pulverized; thereafter floury texture and soft feel	Can form coast that can be handled freely; wet soil runs together and puddles	Cannot form perfect ribbon; has broken surface, cracks appear
Clay loam soil	20-30% clay	Fine-textured soil; quite cloddy but some crumbly material	Lumps hard; not easily broken	Can form cast that can be handled freely; soil plastic	Can form perfect ribbon but breaks easily
Clay soil	30-100% clay	Fine-textured soil; breaks into very hard clods	Lumps very hard; difficult, if not impossible, to break	Can form cast that can be handled freely; soil plastic	Can form ribbon that will support its own weight
Silt soil	At least 80% silt				

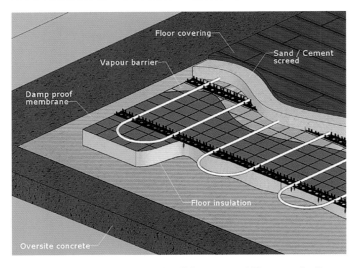

As explained in much more detail from page 147-on, underfloor heating is especially suitable – though not essential – for GSHP systems, because it runs at a lower temperature than conventional radiators. Usually, pipes are installed when a building is first constructed but companies like John Guest/ Speedfit produce fittings that can often be used with existing floor structures.

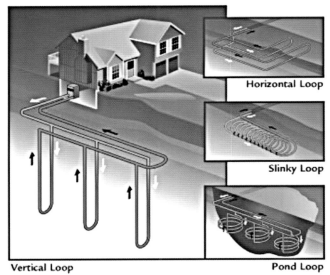

This picture gives an overview of several types of GSHP ground loops. The vertical loop shown here usually consists of one or more bore holes. (Courtesy North American Renewable Energy Directory – www.nared.org)

Even after you've decided that it's both feasible and desirable to fit a GSHP system, there's still a lot of preparation work to be done. Here, from left-to-right: Clive from Ice Energy, Colin Griffiths, the digger man, and Jeremy (Jes) Heywood, our qualified electrician and plumber combined, begin to get to grips with the layout. This was after I had made my initial plans, and was the point where the three experts combined to determine what was and was not possible and desirable.

Because we didn't have room for single pipe runs, we ended up installing a slinky system ...

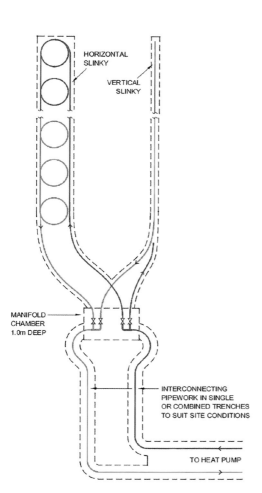

This is an Ice Energy installation of a single pipe run and it gives some idea of the amount of soil that comes out of the ground when digging trenches – and not all of it will go back in again!

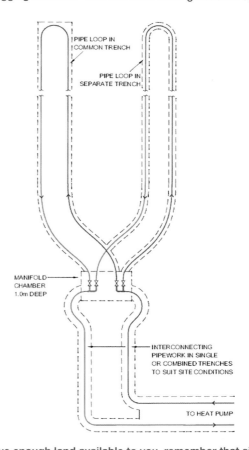

… which consists of pipes part-coiled and laid in the ground either horizontally or in much narrower, vertical trenches, as this drawing from Ice Energy shows.

SECTION THROUGH A HORIZONTAL SLINKY TRENCH

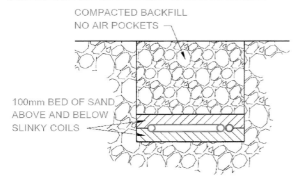

Ideally, your pipe would be buried at least 2m (6.5ft) beneath the surface. As, in most countries, there are health and safety issues with digging to a depth of more than 1.2m (4ft) – check local regulations – in practice, pipes are usually buried at this depth.

If you have enough land available to you, remember that single pipe runs will be more efficient than those overlapping each other. The recommendation is that ideally, there should be at least a metre between pipes and, if they are looping over one another, that clearly isn't possible and the pipes won't extract energy from the ground in the most efficient manner.

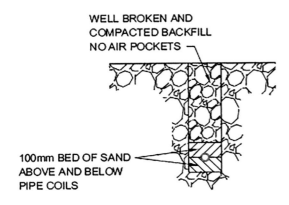

WELL BROKEN AND
COMPACTED BACKFILL
NO AIR POCKETS

100mm BED OF SAND
ABOVE AND BELOW
PIPE COILS

SECTION THROUGH A SEPARATE PIPE LOOP TRENCH

The same back-filling principle applies for single pipes as for slinkies.

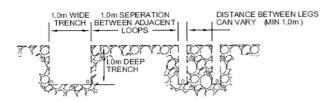

There is an almost infinite variety in the way people choose to lay their pipes. As long as the pipes or pipe runs are the generally accepted 1m (3.3ft) apart, it's just a matter of convenience.

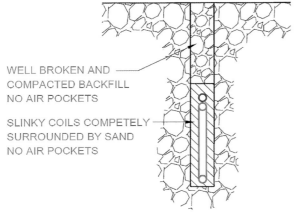

WELL BROKEN AND
COMPACTED BACKFILL
NO AIR POCKETS

SLINKY COILS COMPETELY
SURROUNDED BY SAND
NO AIR POCKETS

SECTION THROUGH A VERTICAL SLINKY TRENCH

One method I have concerns about is the vertical slinky system. If it has been decided that the minimum acceptable depth is 1.2m, most of the pipework in this arrangement will be well above that level. On the other hand, if you were to dig to a greater depth (if that would be permissible under local regulations), it would be difficult to arrange the pipe into coils.

Of mainly historical interest, this is another, very different, system: prefabricated heat exchanger plates buried in the ground. This could be a useful alternative when only a small land area is available, although there is a risk that the heat exchangers won't be anywhere near as efficient at extracting heat as extended pipes because they are closely concentrated. This type of installation is no longer supported by the Microgeneration Certification Scheme (MCS). Specialist advice would be required here. (Courtesy Pedmore Plumbing and Heating)

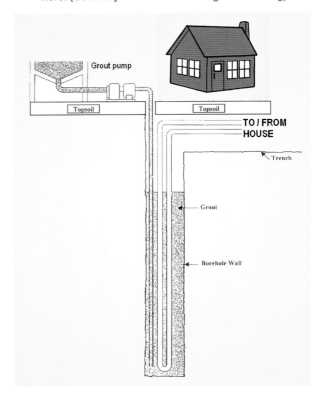

Another alternative is to drill a bore hole, into which a loop of pipe is installed – essentially a flow and return pipe that runs vertically rather than horizontally in the ground. The average depth required in the UK is about 100 metres (328ft), and one major advantage is that ground heat is drawn from a depth at which temperatures are most consistent. Drilling can be a very messy business since water is pumped into the hole as it is drilled to cool and lubricate the drill which creates a good deal of slurry. If this slurry could create a problem, such as in a garden, a good operator should create a bund (wall) to contain it. After drilling, the hole is back-filled with a special grout that sets hard, sealing the bore hole and providing very good thermal conductivity.

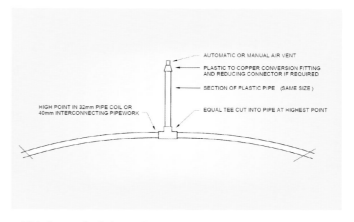

This is a typical air venting solution from Ice Energy for high spots in a ground loop circuit.

Yet another alternative is a water source for which, in the UK, permission is required from the Environment Agency. This can be a lake, pond, river, well, spring or bore hole. Systems are either 'open' loop or 'closed' loop. For the open type, water from the source flows around a heat exchanger, and is then returned to the water source. The 'closed' loop system is similar to a ground-source layout, whereby pipes or heat exchange panels are placed into the water source, which should have a temperature that is never below 8C because, if the heat exchanger freezes, the heat pump will stop working. There are several advantages to this system –

- The CoP (coefficient of performance) and the heat transfer rate from water are usually higher than either the ground or winter-average air.
- The circulation of flowing water provides constant energy replacement.
- The cost of installation can be much less because there's no need to dig trenches in the ground.

The main disadvantages are that very few people have access to suitable stretches of water, while the use of an open loop system necessitates permission from the relevant water authority. The relative vulnerability of pipes must also be a concern where a closed loop system is installed: if the pipework should become damaged, there's a severe risk of pollution caused by the anti-freeze escaping from the system.

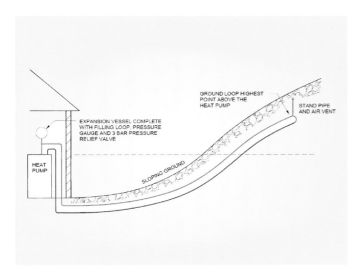

Where a ground or water-source closed loop system is installed and there are any high points in the pipework, they will need to be vented.

Before your specialist draws up plans, it's essential you tell him of any obstructions he may come across. In our case, a mains water pipe runs across the width of the garden. A photograph is as good a way as any to show positions.

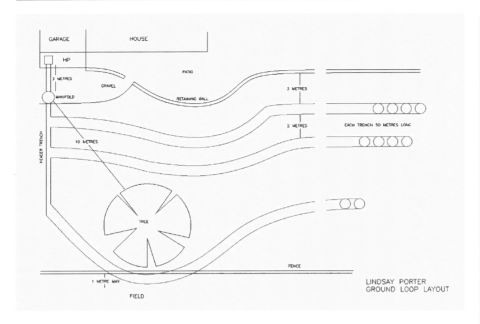

GARAGE · HOUSE · HP · 3 METRES · PATIO · GRAVEL · RETAINING WALL · 3 METRES · MANIFOLD · 2 METRES · EACH TRENCH 50 METRES LONG · HEADER TRENCH · 10 METRES · TREE · FENCE · 1 METRE MAX · FIELD · LINDSAY PORTER GROUND LOOP LAYOUT

As a result of all our planning, this is the diagram that our groundwork contractor used, produced with advice from Ice Energy, and suggesting the lengths and positions of the closed loops in our garden.

The following are the maximum collector hose lengths for IVT's heat pump models when using a ground loop pipe size of 40mm diameter (NB: This is a larger pipe bore than is often used – narrow pipe bores require longer runs – see manufacturer's recommendations.)

Heat pump model	Maximum hose length in one circuit	Maximum hose length per hose in two circuits
Greenline HT Plus C6/E6	600 metres	-
Greenline HT Plus C7/E7	500 metres	1000 metres
Greenline HT Plus C9/E9	400 metres	800 metres
Greenline HTPlus CII/EII	400 metres	800 metres
Greenline HT Plus E14	-	800 metres
Greenline HTPlusE17	-	800 metres

In situations where the length of the collector hose needs to exceed the permitted value, you can connect the hoses in parallel. Note that for parallel connecting the maximum length per hose is specified. As an example the table shows that for E11 maximum hose length is 400 metres. For two hoses connected in parallel the maximum length is 800 metres per hose: ie in total 1600 metres with a parallel connection. For other manufacturers, check their specific data.

IMPORTANT NOTE: Different manufacturers might recommend different pipe diameters. The Ice Energy standard is to use 32mm dia pipe in the trenches and a standard 200 metres (656ft) of pipe per trench in either slinky or straight format. The length of pipe is determined by the ground's thermal conductivity, but a typical 11kW unit in the UK has 3 x 200m pipe loops.

ICE ENERGY PIPE DESIGN CONSIDERATIONS

The minimum permitted bending diameter of collector hose is 1 metre (3.3ft). If sharper bends are required, an elbow connector must be used though only if absolutely necessary because of increased frictional loss. If the collector hose is damaged by too sharp a bend, you can repair the damage using a welded coupling. The maximum length of the collector hose is based on the heat carrier pump's pressure setting. Ice Energy recommends using heat transfer fluid consisting of a maximum of 29% ethanol (by volume) and water. Bio-ethanol has good environmental and technical properties, even at low temperatures, and should therefore be used instead of other heat transfer fluids.

The indoor part of the work on a different installation carried out by Pedmore Plumbing and Heating.

Heat emitters (eg radiators or underfloor heating)

Heat loss calculations for the property must be carried out by your supplier/designer, and a schedule of room-by-room heat losses will have to be issued. Heat emitters will be selected and sized by the installer. The choice of heat emitters in rooms is traditionally between underfloor heating or radiators. The UK's heat emitters are given a 'Temperature Star Rating' which indicates their suitability for operation with a heat pump. The scale is 1 to 6 where emitters with 6 stars will achieve high efficiency (called the Seasonal Performance

Factor – SPF) , and emitters with 1 star will have a relatively low SPF. Full details and a copy of the 'Heat Emitter Guide' for reference can, at the time of writing, be found online at www.microgenerationcertification.org. A quotation from Ice Energy always includes a heat pump selection assuming your choice of heat emitter will achieve at least 4 stars.

UNDERFLOOR HEATING

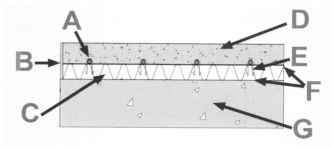

This Speedfit illustration of 'Typical Solid Floor Construction Detail' shows how the finished floor should look. It consists of –

A. Speedfit 15mm B-Pex pipe: laid in pattern shown in design requirements.
B. Edge insulation strip: extends from sub-floor to finish floor level.
C. Floor insulation: must comply with local regulations. Recommended minimum 50mm (2in) rigid foil-faced board for fixing pipe with staples.
D. Floor screed: normally 65-75mm (2.55-3in) from top of insulation.
E. Pipe staples: normally Installed every 400-500mm (15.75-19.7in).
F. Damp proof membranes: normally required on top of and sometimes beneath insulation.
G. Sub-floor: to architect's specification.

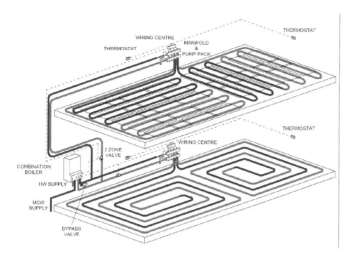

Underfloor heating is generally considered preferable for a ground- or water-source heat pump installation, because a much lower level of heat emission takes place over a much wider area. Some say they dislike having even a low level of heat rising beneath their feet; others don't notice the difference.

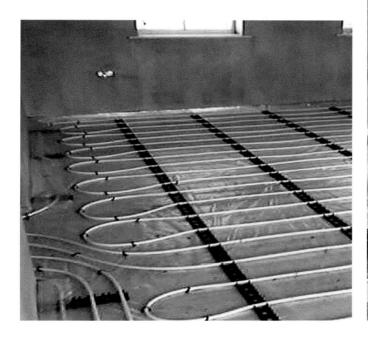

This is a John Guest/Speedfit installation being carried out in a new build. The pipework has been laid out using the Speedfit assembly system before the concrete is laid for the floor.

Other Speedfit systems are available, such as this one for installing underfloor heating pipes to an upstairs room before the ceiling for the room below is fitted.

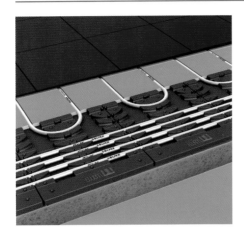

Another approach to adding underfloor heating to an existing structure (as well as new construction applications) is to use Solfex's Warm-Board. This uses a 15mm or 18mm high density, pre-grooved board, carrying the pipes and transferring the heat from the pipes to the floor above, along with a unique (patent-pending) end support system. WarmBoard 10mm or 12mm Pexline 5 layer pipe is laid from the chosen manifold location in a serpentine pattern, and secured within the Warm-Board. The pipe is then supported by the end support with its self-retaining pipe-locking system. The best floor coverings to combine with WarmBoard are hard surfaces, such as stone and tile, as they offer the least resistance to heat transfer compared with carpets. When carpet is fitted, the system requires a higher flow temperature. Specific data is available from Solfex.

IMPORTANT NOTE: When an underfloor heating system is used with a heat pump, it should not be designed so that an entire floor area can be closed down with a thermostat. As much as possible of the circuit should be open circuit. However, in general, thermostatic temperature adjustments for individual rooms are acceptable.

INSTALLING RADIATORS
Radiators can certainly be used with a ground source heat pump system. Conventionally, radiators are sized to allow for the fact that they will be running at temperatures as high as 60°C. Heat pumps are very inefficient when required to heat water to this temperature, but approach peak efficiency when only 40°C is required. In most cases where a heat pump system will be viable – in other

words, in a house with an adequate level of insulation – it is possible to calculate what radiator sizes will heat the house perfectly well with a water temperature of only 40°C. The process of calculating radiator sizes, based on calculating the heat losses for every room, is a little time-consuming but not at all difficult, and there are websites available to help guide you through the process.

Also bear in mind that there are different ways of increasing the efficiency of radiators. In the case of the radiators we chose from Screwfix, most – in addition to being deliberately over-sized – were, in every case, finned versions of single panel radiators, and/or double-panel, finned versions. It's almost always possible to make variations in the pipe spacings to allow fitting of the necessary radiators to achieve required output. What may be a problem in some cases could be a lack of suitable wall area, in which case you may need to look at other types of heat emitter, including those with blown air.

We originally built most of our house back in the 1970s, so took this opportunity for some much-needed refurbishment as well as replacement of the old radiators. A Makita impact driver really speeds up the fitting of radiator brackets.

You can size towel rails in the same way that you size conventional radiators ...

... although you sometimes end up with quite a large one in a relatively small bathroom. (Does mean you can dry a lot of towels, though!) You can buy some pretty exotic and expensive towel rail radiators, but we were very happy with our economically-purchased and very attractive Screwfix rads.

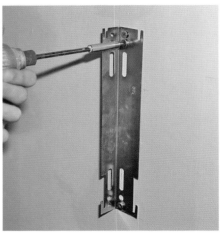

In fact, we ended up purchasing all of our plumbing needs from Screwfix, whose conventional radiators come complete with brackets that are screwed to the wall with wall plugs.

As with most jobs, it pays to spend time planning and checking so that the radiators end up in the correct position – and level!

The easiest job of all is that of hanging the radiator on the brackets.

It's good practice to fit bleed screws and blanking plugs to the top of each radiator as it is fitted, so you don't forget to do so later!

If you're planning to do it yourself, you'll need some plumbing skills, such as knowing that PTFE tape has to be wound in a clockwise direction, while also having a feel for how much to use. The guys from Pedmore Plumbing and Heating …

… are also sufficiently experienced to know that, when breaking into an existing pipe system, there will be some spillage, and an old system such as this one can be quite grubby.

You need to plan in advance which is the best end of the radiator to fit your thermostatic radiator valve (TRV) – an essential part of making the system as efficient as possible.

Ideally, a TRV should be fitted on the flow side, but if this happens to be where it's difficult to get at, or the airflow around it is particularly poor, there's nothing to stop you fitting it on the other side of the radiator.

When fitting resized radiators to an existing system, you'll need to flush out the old system using a proprietary brand of central heating flushing agent.

Then, when the system is recharged, it's equally essential that an inhibitor (or protector) is added to help prevent corrosion in future.

These boxes contained the full set of radiators that we purchased from Screwfix …

ROOM	BTUs	+75%	SCREWFIX NAME	RATING (BTU)
Bathroom (New)	1196	2093	Curved Chrome Towel Radiator 600 x 1500mm Quote No: 76048	1945
Bathroom 1 – Cottage (2 x rads)	2456	4298	Chrome Bar On Bar Towel Radiator 500 x 700mm Quote No: 71573	996
			Double Radiator 300 x 1000mm Quote No: 92590	3398
Bedroom 1 – Cottage	2825	4943	Double Radiator 500 × 1000mm Quote No: 98041	5125
Bedroom 2	1114	1950	Double Radiator 600 × 400mm Quote No: 38412	2361
Bedroom 3	971	1699	Single Radiator 300 × 1000mm Quote No: 83482	1808
Bedroom 4 (2 x rads)	5000	8750	Double Radiator 300 × 1400mm Quote No: 91020	4758
			Double Radiator 300 × 1400mm Quote No: 91020	4758
Dining Room	2721	4761	Double Radiator 600 × 800mm Quote No: 33650	4722
Kitchen	5370	9397	Double Radiator 300 × 1000mm Quote No: 92590	3398
			Double Radiator 300 × 1000mm Quote No: 92590	3398
			Single Radiator 600 × 1200mm CHANGED TO 600 X 1000 Quote No: 36153	3890
Landing inc stairwell	849	1485	Single Radiator 300 × 1000mm Quote No: 83482	1808
Laundry	2455	2455	Double Radiator 600 × 400mm Quote No: 38412	2361
Living Room (3 x rads)	7425	13000	Double Radiator 600 × 800mm Quote No: 33650	4722
			Double Radiator 600 × 800mm Quote No: 33650	4722
			Double Radiator 600 × 800mm Quote No: 33650	4722
TOTAL (btu)	32382 = 9.5kW			

… and their sizes were based on these, the actual calculations that I produced. I'm happy to say that the results were spot-on. Heating throughout the house is gentle with no hotspots. In fact, the house is more comfortable than it was before, with fewer up-and-down fluctuations in air temperature during the day.

GROUNDWORK

The reason this is the largest single section of this manual is quite simple – it is the most important as well as the largest and most time-consuming job covered here. It's essential that you get this right first time because there will be no second chances, short of digging up the entire land area and starting again. And if you get it wrong, your GSHP system won't work properly.

Preparing the ground area

This was the total amount of pipe, in two sizes, supplied by Ice Energy for our installation. High Density Polyethylene (HDP) is the most commonly used type.

Always check the delivery note to make sure that everything is there. Ideally, each item's location when installed should be shown on a plan.

Points to note
- Planning for depth, width and length of pipe runs, as well as how many and where they are installed, MUST be done by those who are qualified and are well-informed in GSHP technology.
- Each pipe circuit should be the same length, so trenches must be planned and dug accordingly.
- The centreline of trenches must be at least 1.5m (5ft) away from boundaries to adjacent properties.
- A standard 50m (164ft) trench will accommodate approximately 30 slinky coils.
- The slinky coils don't have to be

circular; they can be elliptical to make the trench slightly shorter but wider. Bear in mind the need to keep a distance of at least 1m (3.3ft between pipe runs.
- A standard 50m slinky trench will require approx 10m³ of sand.
- A standard 25m (82ft) interconnecting trench (where the pipes are run in a common trench or separate trenches side-by-side) will require approx 2.5m³ of sand.
- Ground loops must be a minimum of 1m away from garden boundaries.
- Remember to take lots of 'witness' photographs during pipe loop installation!

Since this will be a closed loop system, it's essential that absolutely no contamination of any sort is introduced into the pipe. Ensure each pipe ending is properly capped, and if any end caps are missing or loose, use strong waterproof tape to blank off the ends.

Contractor Colin decided to start digging near the house where the closed loop pipes from the manifold would be entering the building.

These pipes would not be quite as deep as the main loop pipes, but that wouldn't matter because they have to be insulated. There's always a mixture of machine and handwork with this sort of job. In the background, you can see the concrete pipes that will be used for location of the manifold.

And here's one of the disadvantages of having clay soil – rainwater is very slow to soak away when holes are left overnight or over the weekend. In fact, the year we chose to install GSHP turned out to be the wettest summer on record, and, throughout the three weeks it took to complete the groundwork, a good deal of use was made of a submersible pump that I'd bought from Screwfix on special offer a year or so earlier on the basis of 'You never know ...'

You can see from the drawing in Part 1: Planning that the plan was for three loops, with a header trench at the top end of the garden. Colin decided to remove the turf and topsoil for the entire width of the header trench ...

...and follow Ice Energy's instructions by running out a single length of the pipe ...

... to the furthest end of the planned first trench.

NOTE: It's recommended that you begin inserting pipe coils at the furthest end of the trench and then, if slightly 'short' of pipe, open up the final coils; if you have too much, you can rearrange them the other way by tightening them a bit – but not so much that coils overlap. On the other hand, this isn't so important if you have more land available and can easily extend your trenches.

The pipe was then brought back to the header trench, allowing it to fall in natural coils as it was laid back down until exactly one third of the pipe supplied for the coils had been used up.

The loops were then measured to make sure that the width was correct for the width of the trench (in reality this may depend on available bucket size of the digging machine) ...

... and a huge number of cable ties were used (not supplied with the pipe in this instance) to keep the slinky pipes in shape. Note that we used my 750kg trailer as a way of moving the pipes to where we needed them.

You may find that you get your loops in a twist the first time you try it ...

... but it quickly becomes second nature. Having confirmed the length of the first trench, the slinky pipe was moved to the other side of the garden.

Then Colin got to work with the digger, lifting the turf and stacking it for use later.

We quickly found that there was room for a longer trench than the stretch of the first slinky. Since you're always looking for greater thermal efficiency, we decided it would be worthwhile redoing the pipe run, making this linkage the same width as it was before, but allowing the maximum distance between loops as could be accommodated in the length of the trench. (When we later received a copy of the Ice Energy manual, the recommendation is that there needs to be a 500mm (20in) gap between each slinky loop.)

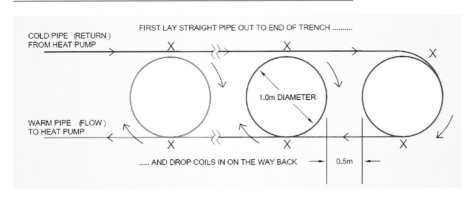

COLD PIPE (RETURN) FROM HEAT PUMP

FIRST LAY STRAIGHT PIPE OUT TO END OF TRENCH ....

1.0m DIAMETER

WARM PIPE (FLOW) TO HEAT PUMP

..... AND DROP COILS IN ON THE WAY BACK → 0.5m ←

The Ice Energy manual recommends laying the pipe in the actual trench. Well, perhaps if you've done it a good many times before this might be the way to go.

The principle having been established, the slinky was moved out of the way once again.

So now we knew how long each trench was going to be – 40m (130ft). Okay, we could have done it by measuring, but groundworkers like to see the evidence with their own eyes, and it certainly is more reassuring.

Work commenced on creating the other two slinky pipes.

There's no getting away from the fact that this is a really time-consuming part of the job …

… but it's important it's done properly so that the actual business of placing the slinky pipes in the trench – a potentially messy business – goes well.

Once Colin began working on the first trench, it became obvious just how much material comes out of an excavated trench. Note the sheets onto which Colin placed the excavated soil to reduce damage to the remaining turf.

Like most experienced digger drivers, Colin is a really accurate worker, though it's still essential to get out of the digger at regular intervals and check actual depth. Some installers use a wooden measuring staff cut to the correct length.

Once the bottom of the trench had been tidied by hand, it was time to lay a layer of sand over the base of the trench. I haven't read this anywhere but decided to use a sticky type of building sand, available from one of our local suppliers, as it seems to me that this is likely to give better thermal conductivity than sharp sand, because the grains of sand are much smaller and the air gaps between them considerably less.

Next, the slinky was lowered into the trench and the pipe pushed down to make sure it was level. If your pipe tends to stick up, you could hold it down with wet sand or even, knock in a wooden peg and tie it down.

If you're working with light soil (not ideal for a GSHP system, it has to be said!) with no bricks or any other hard objects in it, in theory you won't need sand. The sand – and you can see how a layer is being backfilled over the pipe – protects the pipe from sharp objects, and ensures there is good contact between the surface of the pipe and its surroundings: not always easy to achieve when working with clods of clay. However, Ice Energy recommends that you always use sand so there are no air gaps around any parts of the pipe.

Pipe welding

In theory, you shouldn't need to make any joints in your HDPE pipe, although one of several things can happen. You may need to extend the pipe for some reason; you may accidentally damage the pipe, or you may, as in our case, only remember that you have to run the pipe beneath something else *after* you've started backfilling.

Fortunately, HDPE pipe is widely used for domestic, industrial and agricultural water supply, so we brought in the services of Nigel Pitt (NK Pitt Farm Services), a local specialist, who came equipped with his own 110V generator …

… which enabled him to create the 100% secure, fusion-welded butt joint in the pipe.

Nigel kindly left me with a cut-away sample to show how the fusion process works. The connecting piece has a coil within it and electrodes to attach to the generator. Electricity heats the connecting piece, melting the HDPE plastic and making it bond to the connector, creating a butt joint between the two pipes.

Once the pipes were finally laid to rest, so to speak, the remaining sand covering was tipped in …

… and levelled off all the way along the trench.

Incidentally, if you have to work in very wet conditions you may not immediately be able to lay the sand layer on top of the pipe, but at least you'll be able to get the pipe in place and carry out the pressure testing while waiting for conditions to improve.

Note that any fusion-welded joints should, at this stage, be exposed – you'll see why in a moment!

Pressure testing

It is ESSENTIAL that each section of pipe is pressure tested before sand is added, and especially before soil is backfilled. Because of the space constraints we were working with, we were pressure testing and backfilling sections at different times, which is why the sequence shown here may appear out of order. See *How to carry out pressure testing* for details.

Here is one of the many obstructions that you can find in a garden like ours that has been in use for probably hundreds of years: an old land drain crossing the path of the trench. It's not difficult for groundwork specialists to replace sections of land drain after the slinky has been laid.

In a couple of cases, there was evidence of water flowing into the trench but no land drain could be found (it was a wet summer!) ...

... and in those instances, Colin ran new pipes into existing land drains to take away the flow. NOTE: It's ideal if your GSHP pipes are lying in constantly wet conditions. However, if an intermittent flow of water washes sand/soil away from the pipe, leaving no contact, thermal efficiency will be very much reduced.

It is theoretically possible for a GSHP pipe to cause adjacent flowing pipes to freeze. (As explained earlier, although a GSHP system will work less efficiency, it can still extract heat from frozen ground.) In the case of a land drain, it's probably not worth worrying about, but you should certainly take steps if a water main lies in close proximity to the GSHP pipe. I hunted around online and found the cheapest supplier of Armaflex Tuffcote, which is the industry-standard waterproof (closed-cell foam) insulation for use below ground.

It's essential that the insulation is fitted according to the instructions supplied with it. Both sides of the split in the pipe are self-adhesive and protected by pull-off tape.

Here, you can see where the GSHP pipe passes beneath our water main.

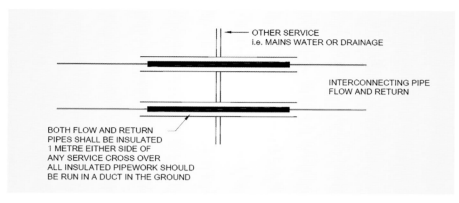

A few rules from Ice Energy –
1. Where ground loop pipes come within one meter of any mains water pipes, sewage pipes, etc, the ground loop pipes need to be insulated 1m each side of the crossing points.
2. For added security the service pipe should also be insulated 1m each side of the ground loop pipes' crossing points.
3. Running services parallel with the pipe loop pipework should be avoided as there is a risk of the service pipes freezing, even if insulation is used.
4. All underground insulated pipework should be run in a duct.

With all of the loops in the ground, Colin concentrated on backfilling the remaining trench.

The oak tree you can see here was an acorn planted by my wife many years ago, so we didn't want to lose it if at all possible. Colin tried to dig his trench no nearer than the outside parameter of the tree's foliage, and we just hoped that that would be enough to spare the roots. (I'm happy to report that Mr Oak Tree is still growing strong!)

Lowering the pipe into the ground in conditions like these isn't the most fun you can have with all your clothes on – but it's better than having to create the coils in the bottom of a sodden trench! It's when conditions are against you like this that you have to be most methodical and determined to do the job properly.

Meanwhile, Tim cut some lengths of 100mm (4in) drainpipe …

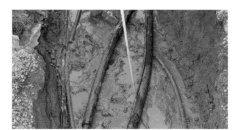

Another place where, if using slinkies, you need to revert to single runs: at the point the pipes enter the common trench. Where different pipes lie in close proximity to one another, it is important to insulate them with Armaflex. It's expensive stuff so Colin is measuring its position.

… for Colin to use as trunking over the insulated sections.

By now, we were back to the header trench, almost where we began. Tim can be seen tipping sand into the final, still slightly soggy, section.

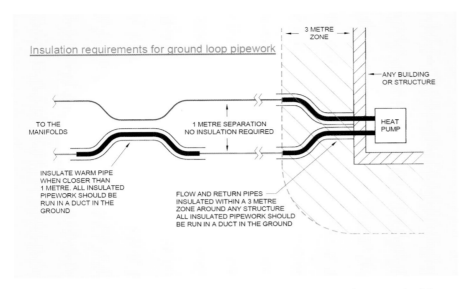

Insulation requirements for ground loop pipework

3 METRE ZONE

ANY BUILDING OR STRUCTURE

HEAT PUMP

TO THE MANIFOLDS

1 METRE SEPARATION NO INSULATION REQUIRED

INSULATE WARM PIPE WHEN CLOSER THAN 1 METRE. ALL INSULATED PIPEWORK SHOULD BE RUN IN A DUCT IN THE GROUND

FLOW AND RETURN PIPES INSULATED WITHIN A 3 METRE ZONE AROUND ANY STRUCTURE ALL INSULATED PIPEWORK SHOULD BE RUN IN A DUCT IN THE GROUND

We followed Ice Energy's instructions for insulating pipework as it approached the manifold.

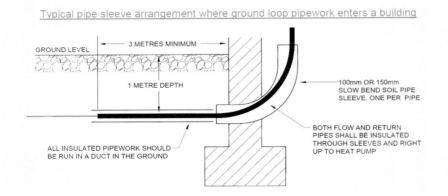

Typical pipe sleeve arrangement where ground loop pipework enters a building

GROUND LEVEL

3 METRES MINIMUM

1 METRE DEPTH

100mm OR 150mm SLOW BEND SOIL PIPE SLEEVE. ONE PER PIPE

BOTH FLOW AND RETURN PIPES SHALL BE INSULATED THROUGH SLEEVES AND RIGHT UP TO HEAT PUMP

ALL INSULATED PIPEWORK SHOULD BE RUN IN A DUCT IN THE GROUND

This is similar to the requirement when passing near a water main, though with subtle differences. For example, this is what you must do where pipework approaches and enters a building from beneath ground level.

How to carry out pressure testing

Now, here's why the pipework itself – as well as any joints in the pipework – should remain uncovered. Jes Heywood put on his plumber's hat and used a hose to fill each slinky loop in turn. It can be quite time-consuming to get all of the air out of the pipe, but it's essential to do so.

Jes then used a pressure gauge to pressurise each loop. While it is possible to air pressure test the pipework (useful if there's a risk of water freezing while waiting for commissioning and the introduction of glycol), Ice Energy DOES NOT recommend air pressure testing – the large stored volume of air within pipework would cause a high energy discharge, and possibly damage or injury should a fitting come loose. All pressure testing should be carried out in accordance with local regulations (see online). Fittings to allow pressure testing were supplied: an end cap for one end of the pipe and an adapter to connect the pipe to a pressure testing pump. The thread on the adapter is a male ½in BSP. You might need to hire a pressure testing pump to carry out this part of the work (available from most tool hire firms). Their connection is normally a female ½in BSP. A common pressure test pump is made by Rothenberger and is used by plumbers for pressure testing of central heating systems.

- In very cold weather, it may be necessary to use glycol anti-freeze, which will be used in the pipework after commissioning. In general, though, because of the health and pollution risks involved when working with glycol (some of which, inevitably, will be spilled when pressure testing), this should be avoided where possible.
- Use the pressure test pump to increase pressure in the loop to 4 bar (take care to follow safety and operating instructions supplied with the pressure tester).
- Make a written note of the start time and start pressure.
- Check and record the pressure every 10 minutes. The pressure will drop, sometimes to about 3 bar. This is quite normal and is because of expansion of the pipe and compression of air in the loop. If there is too much air in the loop, the pressure drop may be greater and you may have to start again, ensuring removal of as much air as possible.
- Once the pressure has stabilised leave for at least an hour and visually inspect the loop to ensure there are no leaks.

- Repeat for each loop. Jes – wisely, in my view – left each loop overnight to make absolutely certain there were no leaks. Ice Energy also recommends you keep the system under test while the trenches are backfilled.
- Any open pipe ends should be sealed as described at the start of this section to prevent dirt ingress. Cleanliness of the pipework is vitally important if running problems are to be avoided in the future.
- Each pipe should be labelled clearly to identify each loop, flow and return.
- Once connected, you must repeat the pressure test for the manifolds and then interconnect pipework from manifold to heat pump.
- Once successfully completed, you may have to sign off a formal Pre–Commissioning Check List.

After each successful pressure test the relevant trenches can be backfilled.

However, it is important to consolidate the soil as thoroughly as possible, which Colin did by having a second machine running with a vibrating soil compressor on the arm. Bearing in mind the pipework and sand added to the trenches, not all of the material you removed will go back in, and will need to be disposed of in some other way.

As one operator backfilled the trench, the other used the consolidator to ensure the soil was packed as tightly as necessary.

The manifold

As we mentioned earlier, Colin used concrete pipe as the manifold chamber. He had to cut access points in the concrete, in what would become the bottom of the chamber when the pipe was lowered into the hole in the ground, and turned the right way up.

Because HDPE pipe is relatively inflexible, Colin had to shorten the pipe lengths to allow him to lower the upper layers of the concrete pipe into position. At the same time, he left enough pipe for Jes, the plumber, to cut to the length he required later on. IMPORTANT NOTE: The only acceptable way of cutting pipe is with a plumbing plastic pipe cutter, such as that shown here, inset. A saw will introduce plastic swarf into the pipe – to be avoided at all costs – and a knife won't achieve the level, square cut required.

Once the manifold chamber had been built up, plumber Jes turned up and laid out, methodically and in order, the components supplied with the kit.

Schematic diagrams showing typical 3 loop flow and return manifold arrangements

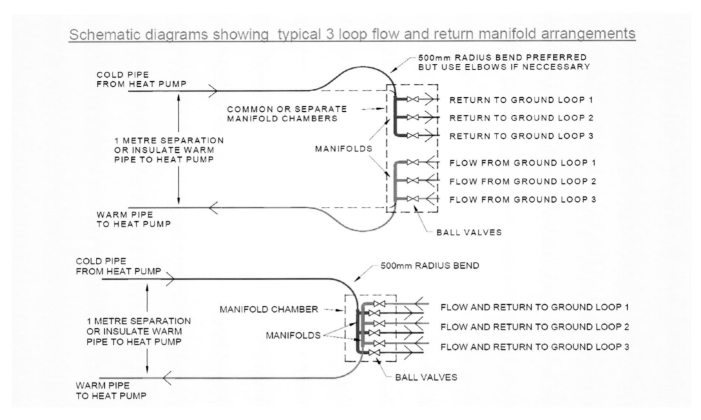

This Ice Energy drawing shows just two potential layouts in the manifold. Whether you place connections end-on-end or on top of each other may depend on how much room you have.

Sensibly, Jes decided to put together any manifold sub-assemblies above ground, where they are easier to work on.

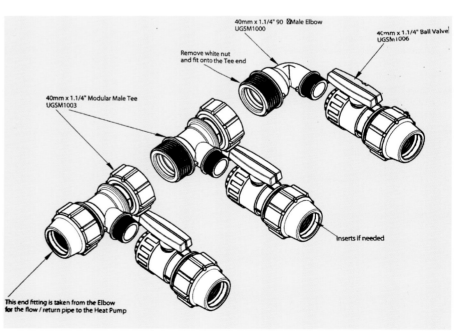

Note: These Plasson ball valves are for 40mm pipe. Therefore, if you have 32mm pipe, use the Plasson reducing set 40-32mm (UGSM1007), which can be inserted into the valves as required.

ASSEMBLY SEQUENCE

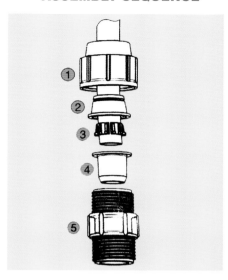

These are component assembly instructions from manufacturer Plasson for the Reducing Set Assembly, 63-20mm –

1) Remove nut (1) from fitting and discard original split ring.
2) Slide nut (1), nut reducer(s) (2) and split ring (3) from set onto the pipe as shown.
3) Make a mark on the pipe to depth of reducing bush (4).
4) Push pipe into bush (4) up to the mark you have made.
5) Push reducing bush (4) and pipe into body of the fitting (5) until the shoulder of the reducing bush reaches the end of the fitting. Some lubricant at this point would help.
6) Draw the split ring (3) and nut reducer(s) (2) close to the body of the fitting.
7) Tighten nut (1) firmly with a wrench.

The bush (4) is fitted onto the tube as described and both are inserted here.

The nut (1) is also fitted to the tube, followed by the nut reducer/s (2) and split ring (3) and the whole assemble is screwed on to the valve, here.

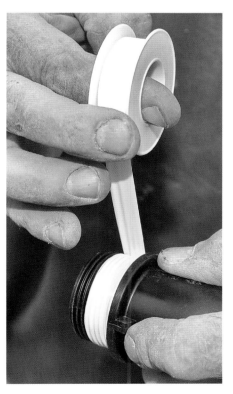

Where Jes needed to add an elbow to a female thread on the manifold, he used plenty of PTFE tape, winding, as always with a right-hand thread, in a clockwise direction …

… using just the right amount of tape so that the fitting would be difficult – though obviously not impossible – to screw on.

If you don't need a wrench to fully tighten the fitting, you haven't used enough tape!

This is a reducer fitting showing how it locates on the end of the HDPE pipe.

But before assembling any of the fittings, Jes used a reamer …

... to ensure that the end of the pipe was perfectly square and free from swarf.

Down in the muddy depths, Jes began to bring together the pipes and manifold sub-assemblies ...

... methodically fitting all of the in-and-out pipes while maintaining perfect cleanliness with all of the fittings.

Obviously, it is crucially important that all of the flow and return pipes are correctly connected, and you need to mark the direction of flow on each pipe. This Jes did with a paint marker, though proprietary marking tape is available.

This is a manifold that does exactly the same job – though with a completely different appearance – which was used on another installation by Pedmore Plumbing and Heating. The point is, there's more than one way of doing the job correctly – it depends on circumstances.

It wasn't until much later – after checks to ensure that the pump inside the building was operating satisfactorily, and that there were absolutely no leaks – that Colin on the digger and Tim on the ground lifted into position the manifold chamber cover.

They backfilled the area around the chamber, and used the Wacker Plate to fully consolidate the soil and stone placed around it.

However, before final backfilling was done, Jes worked on the wall through which the heat pump pipes would pass. It's a sub-ground concrete block wall, so he worked on the outside with a power demolition hammer ...

... matching up the inside with a neatly cut hole.

It was surprising just how much material had to be removed and how long it took. If you find yourself having to go through concrete, don't waste your time: hire a demolition hammer!

The hole was cut to the correct size for a piece of plastic drainpipe to pass through.

In turn, the plastic drainpipe was slid snugly over the pipe from the manifold ...

... which, by now, had been fitted full-length with its insulation.

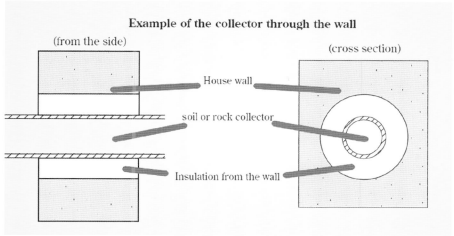

Example of the collector through the wall

(from the side)   (cross section)

House wall

soil or rock collector

Insulation from the wall

These Ice Energy drawings graphically show how to run the collector hose through a wall.

INSTALLING THE GROUND SOURCE SYSTEM HEAT PUMP

Transporting the heat pump
The heat pump should always be transported and stored in an upright position, otherwise the fittings inside the compressor can be damaged. If the heat pump must be tilted during entry to the installation site, this should be for as short a time as possible. The outer cover plates should be removed to avoid damage, if the heat pump is moved without using the supplied pallet. The heat pump must not be stored at temperatures below -10°C.

Installation of the heat pump should be carried out by an experienced plumber, although Ice Energy points out that, unlike the ground loop side, connections to the heat pump require only the conventional plumbing and electrical skills that qualified tradespeople should possess. For that reason, the following are not comprehensive installation instructions, but intended simply as a useful overview of what is involved. In any case, instructions will be specific to whatever make and model of heat pump is used, and will be supplied with the unit.

In particular, your fitter will need to be fully familiar with requirements such as –
• Use of a separate, unvented hot water cylinder, when required.
• Relevant safety kit for an integral

unvented hot water tank, when used.
• Buffer tank and/or a bypass in the heating circuit, as required.
• Fitting of additional pumps installed in the heating circuit, if necessary
• The heating system should be pressure tested and balanced, and corrosion inhibitor added to the heating system prior to your supplier being invited to commission the heat pump.

The installation must be carried out in a manner that meets all standard requirements, and which ensures that the end user is able to correctly and successfully operate and maintain the system.

Positioning the heat pump
It is important that the heat pump stands on a flat surface: adjust the rubber feet so that the heat pump does not lean. There must be a floor drain in the room where the heat pump is sited: this ensures that any water is safely and easily transported away if leakage should occur.

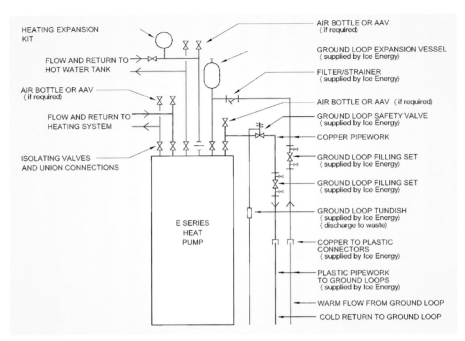

HEATING EXPANSION KIT

FLOW AND RETURN TO HOT WATER TANK

AIR BOTTLE OR AAV (if required)

FLOW AND RETURN TO HEATING SYSTEM

ISOLATING VALVES AND UNION CONNECTIONS

E SERIES HEAT PUMP

AIR BOTTLE OR AAV (if required)

GROUND LOOP EXPANSION VESSEL (supplied by Ice Energy)

FILTER/STRAINER (supplied by Ice Energy)

AIR BOTTLE OR AAV (if required)

GROUND LOOP SAFETY VALVE (supplied by Ice Energy)

COPPER PIPEWORK

GROUND LOOP FILLING SET (supplied by Ice Energy)

GROUND LOOP FILLING SET (supplied by Ice Energy)

GROUND LOOP TUNDISH (supplied by Ice Energy) (discharge to waste)

COPPER TO PLASTIC CONNECTORS (supplied by Ice Energy)

PLASTIC PIPEWORK TO GROUND LOOPS (supplied by Ice Energy)

WARM FLOW FROM GROUND LOOP

COLD RETURN TO GROUND LOOP

Full installation instructions and diagrams will be supplied with the pump you purchase, and suppliers, such as Ice Energy, should be available to provide online or telephone assistance when required.

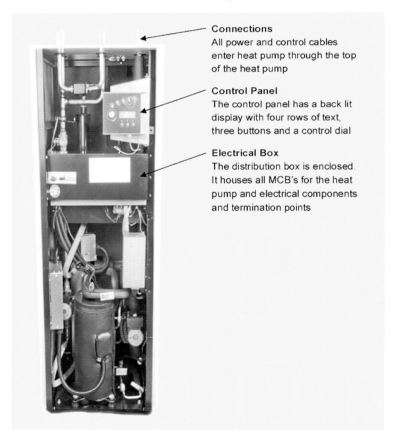

Connections
All power and control cables enter heat pump through the top of the heat pump

Control Panel
The control panel has a back lit display with four rows of text, three buttons and a control dial

Electrical Box
The distribution box is enclosed. It houses all MCB's for the heat pump and electrical components and termination points

This is the basic heat pump layout.

The holes being cut through the wall shown earlier were positioned with location of the heat pump in mind, and with allowance made for pipe runs and ancillary fittings. Note that the pipe insulation must continue through the wall, and the exposed pipe ends must always be closed off.

When making an entry into a building below ground level, there is always a risk of water ingress; particularly pertinent here because floor level at this point is also below ground level. We used expanding foam from a large aerosol can, injected into every part of the access holes, both inside and outside of the plastic drain pipe sleeving. There are also various types and designs of 'puddle flange' designed to help seal a wall against water penetration when a pipe passes through it.

Pieces of wood were used to hold the pipes approximately central in their holes – the bend in the pipe is quite strong – so that the foam could expand right through to the inside of the wall. The foam also provides extra thermal insulation.

Another downpour, another delivery: the Ice Energy heat pump and fittings arrived on schedule. A low doorway would have presented a problem!

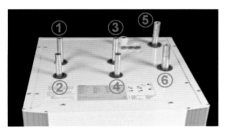

Plumber Jes reconfirmed the pipe in and out locations with the manual.

STORING AND POSITIONING THE HEAT PUMP – ADVICE FROM ICE ENERGY

- The heat pump must always be transported and stored upright. It may be briefly tilted on its back while being manoeuvred into place, but only for a short time (a few minutes), and should never be tilted further than 60° from vertical. Please note that some heat pumps with an internal hot water cylinder are top-heavy, and need to be handled and manoeuvred with extreme care.
- The heat pump is an expensive and delicate piece of equipment and must be protected at all times. It must be kept clean and dry and at a temperature above -10°C at all times, and must NEVER be left outside exposed to the elements.
- Heat pumps can be sited almost anywhere, and while there is no specific ventilation requirement for its location, it is important to ensure that it operates in a surrounding temperature of between 0°C and 35°.
- Prior to final positioning ensure the location is adequate to support the operating weight of the heat pump.
- The main component is a rotating compressor and, although relatively quiet, sound transfer through the building structure may become evident at night-time, so locations close to quiet areas should be avoided.
- Provision should be made for a floor drain or easy means of drainage in the vicinity of the heat pump to allow waste water to be removed in the event of a leak, or during maintenance.
- The weight, dimensions and clearances required around the heat pump will be as detailed by the supplier.

NOTE: Ice Energy points out how important it is that the collector hose has been pressure tested before it is connected to the heat pump. Pressure testing is carried out by the boring company or excavating company, and determines whether the hose is in one piece and works as it should. When laying the ground heat hose the hose is usually pressurised during the laying process. Another leakage test should be performed on the collector hose before commissioning the heat pump.

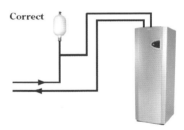

The heat pump should stand on a firm base, preferably concrete, and not a wooden floor. Jes used the rubber feet supplied to adjust the heat pump to a stable, vertical position.

Jes also planned the location of the ancillary components, such as this expansion vessel. A buffer tank, although not required in this installation, will be considerably larger – see page 152.

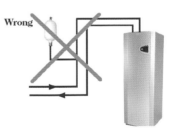

It is important when installing the expansion vessel to position it at the highest point in the circuit, preferably above the heat pump. If a low ceiling makes it impossible to fit the tank above the pump, it can be positioned as illustrated in the upper figure. It is important to install the tank so that any air is dispersed upwards: air will remain in the circuit if the tank is fitted incorrectly (see in the lower figure).

The task of the particle filter is to filter out dirt before it can enter the heat pump. Accordingly, the supplied particle filters should always be fitted on incoming pipes on both hot and cold sides. The filters should be fitted as close to the heat pump as possible, and placed horizontally. Specific details often vary between different designs of heat pump, so check supplier specifications.

The filter is best installed in a horizontal section of pipe to ensure foreign materials stay within the filter thimble when being cleaned. A black polystyrene box fits around the ground loop filling set and filter for insulation purposes, and also contains tools used for regular maintenance of the heat pump, so should NOT be discarded. Special care should be taken while fitting to ensure that this box can be easily removed for future routine maintenance.

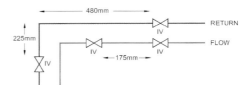

SPACING REQUIREMENTS INCORPORATING A BEND   (not to scale)

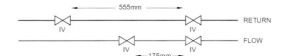

SPACING REQUIREMENTS IN STRAIGHT PIPE   (not to scale)

These are the positioning requirements when fitting full bore isolating valves in the pipework.

The need for a bypass

You can only establish whether or not a bypass will be required via discussion with your specialist supplier, installer and designer. Where the GSHP system includes the heating of domestic hot water (which I don't usually recommend), Ice Energy recommends installation of a bypass as follows –
1. An internal three-port diverting valve switches the flow of hot water from the heat pump to either the heating system or the hot water primary system. Priority is always given to heating the hot water, and when the three-port valve switches over to hot water, no water will flow into the heating circuit.

Usually, all connections are made from the top of the heat pump but, in this case, it was more convenient for one pair to exit from the side of the casing. After discussing the options with the supplier, Jes changed the internal pipework and cut holes in the steel case to allow for the new exit points.

Different heat pump manufacturers may have different requirements – always follow supplier's guidelines – but here's further advice from Ice Energy –

- The heat pump should not be 'hard-piped' (soldered fittings) into the system.
- Isolating valves or unions with compression fittings should be connected on the pipework tails coming out of the heat pump. This will assist with future maintenance.
- Always ensure there are no valves on the pipework between the heat pump and the expansion vessels.
- On the ground loop side, fit plastic to copper connectors and extend the connections in copper to the heat pump.
- Fit the expansion bottle, filling set and strainer on the warm FLOW pipe from the ground loops into the heat pump. and the safely valve on the cold RETURN pipe from the heat pump to the ground loops.
- On the heating side, the connections from the isolating valves extend to the heating system via the bypass circuit or buffer tank, if supplied.

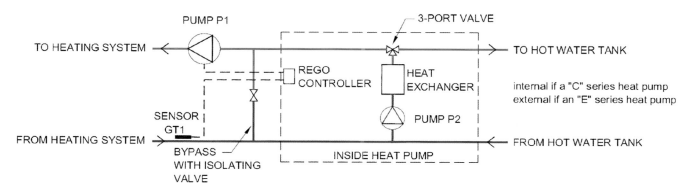

PUMP P1 / 3-PORT VALVE
TO HEATING SYSTEM / TO HOT WATER TANK
REGO CONTROLLER / HEAT EXCHANGER
internal if a "C" series heat pump
external if an "E" series heat pump
SENSOR GT1 / PUMP P2
FROM HEATING SYSTEM / FROM HOT WATER TANK
BYPASS WITH ISOLATING VALVE / INSIDE HEAT PUMP

If a bypass is required (and our installation didn't need one), the pipe size for the bypass needs to be the same as that used for the heating pipes. An isolating valve in the bypass can be used to throttle down the flow rate through the bypass should noise in the heating system become a problem when the heat pump is heating the hot water. Bypass length should be 20x the pipe diameter.

2. The external heating pump(s) (called P1) run continuously when outside air temperature is below the 'Summer Disconnection' temperature to provide the sensor GT1 with an accurate reading for control purposes. (But see pic 15.) When the heat pump is only heating the hot water, a bypass is required to ensure this continuous flow round the heating circuit is maintained.
3. When the hot water system is up to temperature, priority returns to the heating system, and the bypass becomes redundant because all of the water leaving the heat pump is picked up by P1 and pumped around the heating system. The circuit to the hot water primaries remains static until the next call for heat from the hot water tank.
4. The pipe size for the bypass needs to be the same as that used for the heating pipes. An isolating valve in the bypass can be used to throttle back flow rate through the bypass, should noise in the heating system become a problem when the heat pump is heating the hot water. Bypass length should be 20x pipe diameter.

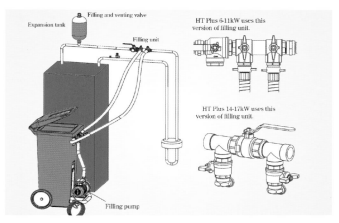

One circuit at a time is filled with heat transfer fluid: keep the valves closed in the other loops during the process. Read more about filling in the specific instructions for the model of heat pump installed if not using the Ice Energy unit shown here.

Water source

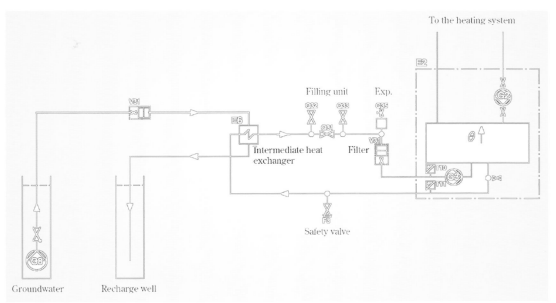

For the small number of water source systems in use, there will be an entirely different way of connecting the heat source pipework from that shown here. There is a lot more to it than this Ice Energy drawing conveys, and specific instructions can be obtained from your components supplier.

Wiring connections

Instructions for the electrical wiring will be provided by the heat pump supplier, and the installation must be carried out by a qualified electrician in accordance with local regulations. We fitted an additional feature – an external timer – because the monitoring system we also installed (see Chapter 1) shows that the heat pump consumes around 80W, even when there is no demand for heat in the house. As a personal decision (and against the advice of some installers), we therefore decided to fit a timer to completely turn off the heat pump at certain times of the night and day. In a well-insulated house such as our own, we found no need for the heating to be on overnight other than in the most exceptionally cold conditions; the sort you don't experience in the UK every winter. The timer works well, though you have to make allowances for the extended start-up and heat-up times compared with fossil fuel, high temperature central heating systems.

The indoor sensor should not be fitted in a room that has any heat-producing equipment in the same room, other than the installed heating system. It should also be located away from anything that will affect the temperature sensed. It is NOT a room thermostat: its job is to give the heat pump controls an indication of how the heating system is performing, so, it should not be fitted in direct sunlight, by a stove, open fire, in line with a draughty door, etc. Any of these locations will give the heat pump false information, resulting in too much or too little heat being produced. Furthermore, the heating system in the room where the sensor is located should not be controlled by a room or radiator thermostat.

Incidentally, when an external switching device (such as a time-switch) is used, the IVT indoor temperature sensor is disabled for the first two hours of the heat pump's operation. Ideally, the sensor should be placed 1.5 metres (5ft) from the floor – this one is awaiting cable extension and repositioning to a different part of the house. The maximum cable length is 30 metres (98ft).

This was the appearance of the almost-completed Ice Energy system after Jeremy had finished installing it. All that remained was to complete the insulation of joints and valves, and correctly label pipe directions.

This is a different installation carried out by Pedmore Plumbing and Heating, showing all of the final insulation and pipe labelling in place. It clearly contains provision for the system to heat domestic hot water – something that tends to make GSHP noticeably less efficient: see the start of this chapter. Insulation standards must be meticulous. Heat pumps work at lower temperatures than traditional heating systems so heat loss to the surroundings must be kept to an absolute minimum. Insulation on the ground loop circuit, where it appears inside the building, must also be adequately vapour-sealed to prevent condensation forming. Ground loop circuit pipework will need to be held away from the wall with insulated blocks and oversized pipe clips to ensure vapour seal continuity. All heating and hot water pipework must be well insulated to ensure the heat gets to where it is required.

Further, vitally important information from Ice Energy includes the following –

• Before fitting the strainer, open and close the valve and unscrew the sealing cap a few times to free up the threads, which will make future maintenance easier. Once fitted, do a dry run to establish that access is acceptable.

• All connections to the heat pump are from above, so there's ample opportunity for air locks to occur. All high points susceptible to air locks must be fitted with automatic air release valves, or manual air vents if preferred.

The following is a summary of all the remaining work recommended by Ice Energy during and following the installation of a ground source heat pump.

IMPORTANT NOTE: Installers should make a note of the volume of water in the primary jacket of the hot water cylinder, which is much larger than traditional cylinders, as this will be required in several of the following stages.

Need for a shunt circuit

If your system has different temperature requirements around the property – for example, a system involving a mixture of radiators and underfloor heating – most likely you will have been supplied with a shunt circuit valve and sensor (the details of this circuit should be provided by your supplier). If you have a requirement for different temperatures, and a shunt circuit valve and sensor have not been provided, bring this to the attention of your supplier/designer at the design stage.

Existing radiators

Any existing radiators, or any other part of an existing heating system that is to be connected to the heat pump, must be thoroughly chemically cleaned and flushed out before connection to the heat pump to prevent any foreign bodies from entering the heat pump. As already noted, the best way to keep heat emitter sizes, especially radiators, to a minimum is to ensure that the building fabric is well insulated, and air infiltration is properly managed to reduce heat loss to a minimum.

Sealed system

Ideally, the heating system should be sealed, but a vented system with a feed and expansion tank is allowed if absolutely necessary, such as in our installation where a wood burner with back boiler was installed.

If a sealed system is being installed, a standard Robo kit expansion vessel with pressure relief valve, tundish, gauge and filling loop need to be sized and installed on the RETURN connection to the heat pump. The size of this vessel is dependent on the size of the system, and must be calculated by the installer.

Pressure testing the heating system

Once installation of the heating system is complete, the entire system must be pressure tested to twice the working pressure: eg if standard working pressure is 1.5 bar, test pressure should be 3 bar.

Corrosion inhibitor

The installer is responsible for adding inhibitor to the heating system, in an amount commensurate with system size.

Commissioning the heating system

When the heating system has been flushed and the inhibitor added, it should be fully commissioned by the installer to the requirements of the currently applicable specifications of the region in which the installation has been carried out.

Extra points to note

- All high points in pipework which are likely to air lock must be fitted with automatic air vents, or manual air vents if preferred.
- All pipework must be identified with the service it is carrying, and direction of flow must be indicated.
- Discharge from the ground loop safety valve should flow into a suitable container so that any liquid can be returned to the ground loop expansion vessel if necessary.
- If locating the ground loop filter on the heat pump side of the expansion bottle with a downward vertical flow is the only place where it is accessible, the isolating valve on this pipe can be omitted.

Monitoring details

Depending on location and associated regulations, it may be necessary to fit monitoring equipment for when financial incentives become available from government. Virtually all heat pumps are designed with this in mind: for instance, Ice Energy's E-Series heat pumps have provision for two sets of monitoring equipment – one on the heating system and the other on the hot water primary system. In this instance, each set of monitoring equipment comprises a flow meter and temperature sensor located in the RETURN pipe to the heat pump, and a temperature sensor in the FLOW pipe from the heat pump.

Installers must install full bore isolating valves in the pipework to these systems with adequate space between to enable monitoring equipment to be installed as detailed below, and in a location that will enable monitoring equipment to be easily retrofitted, without draining the system. The elements of each monitoring set must be within 2 metres (6.5ft) of each other to enable correct wiring.

ELECTRICAL INSTALLATION – GENERAL NOTES
Introduction

All electrical works must be installed by a suitably qualified person in accordance with local, current regulations, and all power connections to the heat pump must be carried out by the installer. All control elements – such as sensors, valve motors, etc – should be fitted by the installer, and the wires run back to the heat pump. These wires must be clearly identified and coiled neatly for connection by the supplier's engineer (in our case, Ice Energy) during commissioning.

Incoming mains supply

In all cases, wiring and electrical protection systems must be in accordance with local, current regulations. The following can only be considered a general guide, and does not take precedence over any applicable regulations.

There must be a separate, motor-rated circuit breaker (C-rated) or fuse; a local isolator must be installed adjacent to the heat pump. Requirements for RCD protection and earth bonding is that they are installed in accordance with current regulations. The use of a 30ma RCD is not recommended due to nuisance tripping. A 100ma RCD for protection against fire would be suitable. Adequately-sized cable must be installed from the isolator to the terminals in the heat pump: cable size to be be determined by the installer (usually dictated by the breaker/fuse rating and distance from the local isolator). Electrical loadings of the heat pump should be given in the data information supplied with the pump.

INTEGRATION OF UNDERFLOOR HEATING CONTROLS WITH THE HEAT PUMP¶

With buffer vessel

If the underfloor heating manifolds are fitted with circulating pumps, and there is a buffer vessel in the system, these manifold pumps can be controlled from the underfloor heating wiring centre, if preferred.

Without buffer vessel

If there is not a buffer vessel in the system the manifold pumps DO need to be controlled by the heat pump, and these should be wired back for connection to the heat pump because, in these circumstances, at least one underfloor heating circuit must be open with no thermostat control to allow the pumps to circulate water around an open system. This is essential for the heat pump controls to function properly, and will ensure the pumps are not pumping against a closed head. A relay may be required to control all of the pumps.

CONNECTING TO EXTERNAL (P1) PUMPS

All heating system pumps external to the heat pump are called 'P1.' In a standard arrangement, where either a bypass circuit or buffer tank have been installed, one or more external heating pumps will be required. The connection method of external pumps may vary between manufacturers, but an example of requirements for the Ice Energy system being installed follows.

On simple installations, there will be one or two P1 pumps. The P1 pump or pumps can be connected directly into the P1 terminal on the printed circuit board located in the

heat pump's electrical distribution box, provided that –
- The pump/s are located in the same room as the heat pump.
- The total current consumption of all pumps does not exceed 6A.
- A maximum of two P1 pumps are used.

If the pump/s are located in a different room from the heat pump, if the total current is likely to exceed 6A, and/or there are more than two pumps, a relay must be installed, the coil for which should be wired to the live and neutral terminals of P1 in the heat pump while the power to the heating pumps should be from a separate, fused mains supply.

CONNECTING TO THE MAINS SUPPLY

Mains power supply to the heat pump is connected via the installer into the Live, Neutral and Earth connections on the internal terminal blocks located in the electrical box.

TEMPERATURE SENSORS

Again, sensors for different heat pumps may vary, but these are typical.

Outdoor temperature sensor

Typically, this sensor requires a 2-core cable, 0.5mm^2 from heat pump to the position of the outside temperature sensor. It should be sited on a north-facing wall in the northern hemisphere, out of direct sunlight: under-eaves is often ideal.

Indoor temperature sensor

Some heat pumps will work without an indoor temperature sensor, though this is not advisable. This is NOT a room thermostat – it supplies extra data to the heat pump's microprocessor but does not, by itself, turn the heat pump on or off. Typically, this sensor requires a 0.5mm^2 4-core cable from the heat pump to the location of the indoor temperature sensor. It is recommended this be sited in a room with open-zone heating (ie not separately controlled by thermostat), where it will not be influenced by any other heat source such as a wood burner, Aga, etc, at a height of 1.5m (5ft) from the floor.

Heating system return water sensor

This sensor is attached via cable ties to the return pipe from the heating system, inside the insulation, and adjacent to the heat pump, and its connecting wire is run back to the heat pump.

Domestic hot water temperature sensor

When domestic hot water is included with the heat pump, this sensor should be installed in the hot water cylinder used for your domestic installation. The cable should be run back to the heat pump, and connected in accordance with manufacturer's instructions.

Additional sensors

Different sytems may require separate sensors to those described here, depending on the type of system and any additional items supplied with your heat pump.

IMPORTANT NOTE: If sensor cables run alongside mains power wiring, the sensor cable should be screened.

Maintenance

One of the great things about a ground source heat pump system is that, unlike several other renewable technologies, it's almost maintenance-free! Clean out or change the filters (according to model) at the intervals recommended by the supplier; visually check a couple of other items – and that's it! Ice Energy recommends that you check the following items a few times during the first year, and then once or twice a year thereafter –
- Sight glass.
- Expansion vessel.
- Particle filter.
- Protective anode (only applies to installations with a stainless steel hot water cylinder).
- The system pressure should be checked, using the pressure gauge incorporated within the system (see individual manufacturer instructions) as being in the range 1-1.5 bar, to prevent unwanted water ingress from the mains water supply.

The user manual supplied with your heat pump will tell you where the sight glass is located. Sometimes when the heat pump has started you can see the fluid in the refrigerant circuit bubble for a few minutes in the sight glass – this is completely normal. However, if it bubbles continuously you should contact your dealer. If the sight glass shows green this means there is no moisture in the system; if it is yellow there is moisture in the system. If the latter happens, contact your dealer.

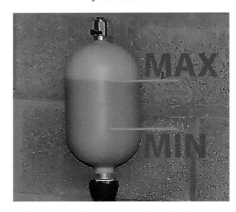

On the Ice Energy Greenline HT Plus, a plastic expansion tank is connected to the heat pump heat transfer circuit (cold side), enabling you to see the fluid level through the plastic. The level in the tank should not fall below the bottom third of the tank. If the fluid level is too low, contact your dealer. If the level has dropped too low, after agreeing with your dealer that there are no existing faults in the system, filling can take place as follows (the heat pump must be operational during filling) –
1. Remove the cover on the valve on top of the tank and open the valve.
3. Pour in anti-freeze, or ant-ifreeze diluted with water to the recommended ratio, until the tank is approximately two thirds full.
4. Close the valve and screw the cover back on.

Filterball® servicing

This is how the filter in the superb Marflow Hydronics Filterbal® (a combined isolation valve and strainer unit) is serviced. The inbuilt strainer screen removes the need for separate isolating valves. The Filterball® comes supplied and fitted with a standard 700 micron strainer (0.7mm hole size), which is interchangeable with a range of strainers from 180 micron (very fine) to 800 micron (coarse) as factory-fitted options or as interchangeable spares. Your supplier should supply you with the correct one, at advise on the correct strainer, if you are supplying your own.

The Ice Energy kit for the ground loop fluid includes a valve and filter pack, which should be contained in the insulated packaging in which it is supplied. This is NOT disposable packaging – it's intended for use, as shown here. Perhaps I'm a bit slow on the uptake but it took me a while to work out what the components were – so I labelled the insulation to avoid further confusion! The valves are for filling the system.

As you unscrew the cap and close the valve, a small amount of fluid (about 0.5 litre in my case) will be lost, so have a container ready to catch it.

- The fluid reservoir will drop by this amount but, if you use a scrupulously clean container and filter the fluid before it goes back in, there's no reason why you shouldn't top up with the fluid that you inadvertently drained.
- The fluid contains anti-freeze which can be both attractive and harmful to small children and animals. Be sure to keep the drained fluid safely covered, dispose of unwanted fluid responsibly, and thoroughly wash out containers after use.

Shut down the heat pump using the ON/OFF button. You'll need a large, adjustable spanner or Stillsons to unscrew the seal nut. Inside, you'll see the ball which ...

... when rotated with the detachable handle both exposes the filter and seals the pipes.

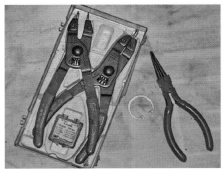

The Marflow Hydronics Filterball marine grade, stainless steel strainer screen is enclosed within the ball, and is accessed for cleaning or maintenance when the ball is in the isolation mode. (Courtesy © Marflow Hydronics)

Supplied with my kit were a pair of long-nosed pliers (right) for removing the circlip holding the filter thimble in position: not completely useless but close! Purpose-made circlip pliers, like the Sykes-Pickavant set I've been using for years (left), are vastly superior ...

... because the shaped ends grip the holes in the circlip under lightly-sprung force. (Of course, you may love the sometimes lengthy pursuit of Hunt The Circlip. It's a game the whole family can play ...) From beneath the circlip, the filter thimble can be extracted.

Note that, after cleaning the filter and when reassembling, the concave side of the filter should face the direction from which the fluid is flowing so that it can contain any debris. It is possible to turn the ball in either direction when opening the ball valve again. I added a direction arrow to the filter housing.

The filter for the domestic heating fluid is inside the heat pump casing. It's a similar setup and, while it is possible to remove the circlip with the pliers supplied, replacing it is a lot trickier and I don't recommend it.

The Renewable Energy Home Handbook

In this case – probably because the old central heating system had been so extensively refurbished – there was a lot of debris in the filter the first time it was checked. If fitted the right way round, it fills from the inside, so there's a lot here trying to poke its way through the mesh!

The cleaned filter thimble was refitted and the valve reopened, as for the external valve described earlier.

Checking the protective anode

This only applies to heat pumps with an integrated, stainless-steel hot water cylinder. A protective anode is located in the top of the cylinder with the task of preventing corrosion. The hot water cylinder must be filled with water in order for the anode to work. On IVT Greenline models, a lamp inside the front cover indicates the anode status. If green, the protective anode is operating and working normally. Red can indicate a fault. When large amounts of hot water are used (eg when running a bath), the lamp may show red for a short period without there being a fault. However, if the light is on for more than ten hours this indicates that the anode is faulty, and you should contact your dealer. If the fault occurs at the weekend you can wait to the next working day before contacting your dealer.

Checking the system pressure

1. Locate the heating system pressure vessel which is usually bright red. Check the black pointer on the pressure gauge: this is the active one. The red pointer is adjustable with a screwdriver to give the user a reference point.
2. If in the range 1 to 1.25 or 1.5 bar (check specifications), no further action is required.
3. If below 1 bar, look for the braided hose from the pressure vessel to the mains water supply.
4. There is a ¼ turn valve at each end of the braided hose. These are 'OFF' if the tap is perpendicular to the pipe. If in line with the pipe they are 'OPEN.' They should both be 'OFF.'
5. To top up the system, open one tap then, while looking at the gauge, carefully open the other tap, and watch the gauge as it rises. When in the range 1-1.5bar, turn both valves to 'OFF.'
6. If over 1.5 bar, chances are the taps have been left open. Close them and reduce the pressure in the heating system by twisting the pressure relief valve.
7. If it is necessary to keep topping up the system there's a chance there's a leak somewhere in the system which should be found and fixed.

PART 5: DRILLING BORE HOLES

A different approach to that of laying pipes along the ground is to drill deep into the ground in order to tap into the thermal energy accumulated there.

This is NOT the same as geothermal energy, which is produced after drilling to a far greater depth, and extracting heat from the still-hot Earth's mantle (in most of the world, except for some volcanic regions where the mantle's heat is near the surface). Using a bore hole to supply heat for your ground source heat pump still uses heat that has been absorbed from the sun – it's just a lot further down and has a more consistent temperature.

Much of the following is taken from an Ice Energy publication entitled, 'Bore holes for Heat Pumps – Specification and Location.' Ice Energy states that heat pumps must be designed and specified to provide sufficient energy collection from the ground to allow its ground source heat pump to operate efficiently. If the chosen source is a bore hole, Ice Energy ensures that the bore hole is always created according to the manufacturer's criteria and (we only work with drilling companies who are affiliated to the British Drilling Association). It's a good idea for anyone commissioning their own work, to use a company recommended by a recognised trade body.

The depth of the bore hole will be dependent on several variables, including the heat load of the property and the geology of the local ground, both of which will have been established during the project specification and site survey.

This is a rig used to drill bore holes for heat pumps. Ice Energy operates several rigs of varying sizes, and can accommodate drilling bore holes for heat pumps in the smallest of gardens, right up to the biggest building sites. The company uses a conventional rotary system, whereby the rig rotates drilling rods into the ground, cutting a way in. This method is relatively clean, although there will be some spoil and water discharge (always tidied up by the company's on-site team): something that should be pre-checked with any installer.

One Canadian contributor to a green-build forum recently wrote, "We didn't contract the clear-up of the drilling debris (due to ignorance), and ended up shifting about 20 barrow-loads of gravel-sized rock [and] the alley between us and the lower neighbor ... got covered with a slick of clay from the drilling."

How many holes; how deep, and how big?

The specification of bore holes for heat pumps is calculated using software that takes into consideration the size of the heat pump, the geology of the ground, and the heat losses of the house or building. Bore hole installations will usually be between 60 and 120 metres (196 and 394ft) deep, and may consist of between one and 12 bore holes depending on project size. A standard bore hole is approximately 120-150mm (4.7-5.9in) in diameter. Using the foregoing method, Ice Energy expects to drill between 50 and 60 metres (164-197ft) of hole per day.

What goes in the bore hole?

Once the drilling of bore holes for heat pumps is completed to depth, the bore hole loop is inserted in the hole. This loop consists of a pair of plastic pipes, joined together at the bottom

Here's an example of a site after GSHP bore hole drilling has been completed and tidied up. As soon as the bore hole loop has passed the pressure test, the hole is backfilled. It is important that the backfill material is the correct type, and that it fills the hole completely. Some recommend that the specialist should pump in a natural clay compound, mixed with water, into the bottom of the hole, which solidifies over time. When finished, all that remains will be the tails of the pipe protruding approximately 500mm (19.6in) above the ground. The drillers will attach end taps to the pipes to protect them, leaving the bore holes ready for connection to the manifold.

by a 'U' bend. (This bend is joined by a method called electro-fusion welding and is manufactured and tested in the factory.) Once the loop has been lowered into the hole it must be filled with water and pressure tested.

It's worth noting that, when the bore hole is drilled, quite a lot of spoil comes out. Also, when the drill is operating, it is not safe to stand nearby. Soil, stones and perhaps water can fly out up onto the air, up to a distance of 5 metres (15ft) – and sideways, of course.

Bore hole drilling practice

The following photographs and information were compiled by Richard Lane, Managing Director of Geologic, one of the UK's leading specialists in drilling bore holes, and explain what would happen if you arranged for a specialist to drill a bore hole in your garden. Geologic says it can use a 'MudPuppy' mud handling system, which allows the drill cuttings to be separated from the drilling mud, and deposited into a skip, reducing the amount of slurry and debris and its spread.

A selection of types of drilling rig. The three on the left are rotary, top-drive, tracked, and hydraulic drill rigs, mainly used for drilling deep bore holes through any formation, including hard bedrock. The red triangular rig is a cable percussion, or 'Shell & Auger' rig, mainly used on geotechnical investigations (though Geologic has drilled shallow GHSP bore holes with these rigs), and the rig on the right is a small, tracked, rotary/window sampling rig, predominantly used for site investigations.

This is one of the rotary drilling rigs, being used in a domestic location, ready to drill bore holes for ground source heat pumps on a small, domestic, social housing project. The bore hole was drilled using mud flush, where drilling fluids known as polymers are used to bring drill cuttings to the surface, while drilling takes place through a mud handling system that allows clean drilling. This prevents drill cuttings being bought to the surface under pressure, and allows the site to be left tidy after drilling has been completed. This is the preferred system when drilling in domestic residential situations.

This rotary drilling rig is being used on Dartmoor in Devon where the ground is very hard granite. It's another GSHP being installed at a single domestic property. The air-flush method (instead of mud-flush) is being used. Some geological formations are easier, quicker and cheaper to drill using traditional air flush, especially in very hard rock formations such as this one.

This is one of Geologic's top drive, rotary-tracked drilling rigs, which is capable of drilling to approximately 120 metres (400ft) below ground level. Weighing in at around 4 tonnes it is still very manoeuvrable, and with a width of only 1.2m (4ft), is perfect for accessing residential gardens and narrow-access sites.

A steel casing that has been drilled into the ground for the first six metres (20ft) of the bore hole. This allows any superficial deposits to be supported during drilling, and prevents them collapsing: essential in bore hole drilling, says Geologic, because collapse risks the drilling string (sections of steel rods) becoming jammed in the ground, and would also prevent the installation from being successfully inserted into the bore hole.

One of Geologic's 10 tonne rigs drilling shallow piles for the installation of steel supports on a solar farm. It's not GSHP, but it just shows how bore hole drilling can have a useful place in other renewable installations, too. This rig has a fully articulating drilling arm which allows straight holes to be installed on variable and sloping ground such as this.

Whoosh! – a large water strike during drilling. This rig is drilling for water on a large farm for a commercial, private water supply – a similar process though with a different end in mind. The drill cuttings and spoil are bought to the surface using compressed air which, in turn, brings water to the surface in a fairly spectacular way!

HEAT PUMP TRICKERY REVISITED

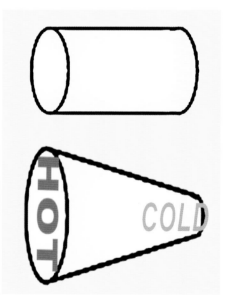

So, how does a heat pump take a small amount of heat and turn it into a larger amount?

Imagine you have, on your lap in front of you, a piece of modelling clay. It's a round bar, about, oooh, THIS wide and with parallel sides –

like a good ol' rolling pin. The parallel sides are going to represent, in this imaginary trick, the outside air temperature. We'll call the left-hand end of the bar HOT and the right-hand end COLD. Now, re-shape that bar, making it into a long cone – much fatter at the HOT end and much narrower at the COLD end.

Okay, so now you've moulded the clay (remember, it 'was' the outside air temperature) so that it is larger at one end, and a corresponding amount smaller at the other end. Same amount of clay; same amount of 'temperature' – it's just been squeezed so one end is 'HOTTER' than it was before and the other end is 'COLDER.'

Well, that's what a heat pump does, and the above is a great analogy. The heat pump 'squeezes' (ie compresses) air or fluid at whatever temperature it starts off at, making the air/fluid HOT at one end and COLD at the other. The heat is injected into your house and the cold is sent back outside again. (Or vice versa if the heat pump is running air conditioning.)

Of course, squeezing the clay took the energy of your hands (or at least, the energy of your brain thinking about that virtual piece of clay!). Compressing air or fluid also takes energy: the electricity used to run the heat pump and compressor. But because it's taking heat from an existing source, it's typically creating three times as much (heat) energy as the (electrical) energy it consumes. It's not magic – but it's a damned clever trick, all the same!

Chapter 9
More renewable technologies

There are so many other types of renewable energy – with more appearing all the time. There are even combinations of existing technologies, where solar thermal and solar PV are combined so that the solar thermal process is said to remove excess heat from solar PV panels and make them operate more efficiently. Moves are afoot to harness body heat to power tiny objects attached to clothing, and for body heat generated in mass transport systems to be captured and reused. With long-term increases in fuel costs and the need to reduce fossil fuel use, expect more to appear as time goes by. In the meantime, the following are some of the more common 'alternative' alternative energy sources.

DOMESTIC TECHNOLOGIES
- Exercise bike generator.
- Fuel cells.
- Heat recovery.
- Micro-CHP (micro combined heat and power).
- Micro hydro power.
- SVO (straight vegetable oil) generator.

LARGE-SCALE TECHNOLOGIES
- Anaerobic digestion.
- Wave power.

The following sections on anaerobic digestion, micro hydro and wave power were all adapted from information published by DARE (Devon Association for Renewable Energy).

Exercise bike generator

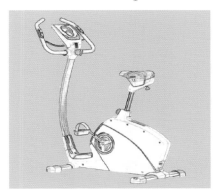

These devices undoubtedly generate more heat and energy as a topic at middle class dinner parties than they do from pedal power. The reality is that producing power from an exercise bike might be fun, but it would be neither cost- nor time-efficient. The bike sketched here generates only enough power for its own on-board lights and metering system, bringing a whole new meaning to the expression 'exercise in futility.' One Danish hotel which has exercise bike generators connected to its electricity supply estimates that one guest cycling at an average speed of 20 miles per hour for a full hour will produce approximately 0.1kWh. So, even allowing for over-optimistic calculations, have you got the stamina or the time to spend 10 hours cycling at 20mph in order to generate 1kW of electricity? And don't forget the expense of buying or building the thing in the first place. On the other hand, if a bank of cycle generators could be harnessed in fitness clubs, where they may be used for hours at a time, perhaps the money and effort invested could begin to pay off. If not, perhaps a device for capturing hot air at dinner parties would be more efficient …

Fuel cells

As this Baxi illustration shows, fuel cell micro-CHP works by taking hydrogen and using it to produce electricity (and heat – see CHP). The hydrogen is passed over the fuel cell where it reacts with oxygen in the air to produce water and energy: 2 x Hydrogen + 1 x Oxygen = Water + Energy. Theoretically, the world's supply of hydrogen is inexhaustible and the pollution effects from fuel cell are nil. So, why isn't fuel cell technology everywhere from houses to cars and factories? The first reason is cost: each cell requires a relatively large amount of platinum as a catalyst; the second is the cost of producing hydrogen, and the third includes the cost, risk and infrastructure required for storing and transporting the gas. For now, the technology is still at the developmental stage and not widely available to consumers. One day, perhaps … (Energy Saving Trust)

Heat recovery

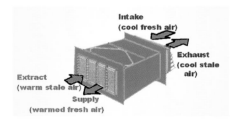

Heat recovery systems extract as much heat as possible from warm air that is exiting a room or a building, and use the heat to warm the fresh air being drawn in. The simplest and least expensive form of heat recovery unit is one that replaces an extractor fan, and recovers heat only in its immediate area.

Heat recovery systems can also operate on warm waste water, where energy usually flows down the drain – literally and metaphorically.

In new buildings built to best ventilation standards, it is much more practicable to recycle heat from extracted air and so-called mechanical ventilation with heat recovery (MVHR) systems (also known as 'whole house ventilation' systems), which are sometimes fitted.

Heat recovery systems often have concealed ducts in ceiling voids leading to the heat exchanger, which is tucked away in the loft or other suitable space. Some have boost settings which can help extract excessive moisture from kitchens and bathrooms. Air filtration is frequently included, preventing pollen and other particles getting in, to the benefit of allergy sufferers. Systems usually run continuously and are normally said to be inaudible.

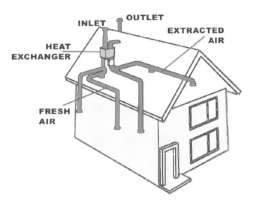

The Carbon Trust describes a typical installation. When cool, fresh air is drawn into the system, it passes over a heat exchanger which captures heat from the outgoing air. Some of this heat is transferred back into the cool, incoming air, reducing the amount of heat that the building's heating system has to generate. The recovery efficiency of this type of system, it's claimed is typically between 55 and 65 per cent. It is also possible for a heat recovery system to move warmer air from one part of the house (such as a south-facing conservatory or room with log burner) into cooler parts of the house. At the same time, unwanted air from cooking, bathing and toilet areas is automatically extracted while as much heat as possible is retained from these (relatively) warm areas.

Micro CHP (combined heat & power)

This technology generates heat and electricity simultaneously from the same energy source, in homes and buildings, individually or in groups. Traditionally, CHP systems have been used for district heating and commercial applications because of the greater inefficiencies involved. But CHP can and does also work well on the domestic scale.
(Illustration EST)

An off-Grid home with a substantial generator – especially one that is water-cooled – can be connected so that waste heat from the engine and exhaust provide hot water to the house.

The award-winning Baxi Ecogen looks like a conventional wall-hung boiler, and (a) uses a conventional gas boiler burner, connected to the mains gas, to heat helium. The helium (b) in the hermetically-sealed Stirling engine* expands and depresses a piston. Cold water flowing around the boiler absorbs the heat, the gas contracts, and the piston rises again. The heated water flows out of the boiler and into the hot water cylinder. Cold water flows into the engine and the process begins again – the piston is driven up and down 50 times a second. The piston has a magnet attached to it and, as the magnetic field passes through a coil at the bottom of the engine, it generates up to 1kW of electricity (c). Much of the heat expelled in the exhaust gases from the system is captured by a heat exchanger (d) and reused for generating domestic hot water. Servicing costs and maintenance are estimated to be similar to a standard boiler – although a specialist will be required.

The Stirling engine was patented in 1816 by Robert Stirling, a Scottish clergyman.

SVO (straight vegetable oil) ATG generator

If you intend running off-grid, or if you want to be certain of having a back-up supply, you'll need a generator. One of the biggest downsides to using a 'genny' is the cost of running it. Petrol generators cost most to run, with LPG and diesel costing less. But one of the greenest ways of using a generator is to run it on vegetable oil power. It's sometimes possible to obtain used vegetable oil at no cost, and if you run your own catering business, you will almost certainly

have access to it. (Strictly speaking, used oil should be disposed of by a licensed company.)

Used veg oil needs to rest to allow the impurities and solid fats to settle at the bottom, and is then filtered to at least 5 microns so that it doesn't clog the injection pump on the diesel engine. Some people invest in a pump and oil filtration kit; others allow gravity to (slowly) do the work using filter socks. There's loads more information online if you search under 'filtering waste vegetable oil'.

The diesel generator's engine must be prepared so that it can be started on diesel, and then switched to veg oil when the engine has warmed up. This is so that the oil becomes warm enough and thus thin enough to run the engine. Then, when the engine is to be turned off, you need to be able to switch back to diesel so that the injection pump is clear of vegetable oil (which will go thick when it cools), and primed with diesel again, ready for the next start-up.

You also need to be certain that the generator's diesel engine is okay for use with vegetable oil. It's certainly the case that the more sophisticated car diesel engines, with very high pressure, common rail injection systems are not suitable for use with vegetable oil. On the other hand, old-fashioned engines, such as Lister engines fitted to generators, apparently use vegetable oil quite successfully, once they've been adapted to take a twin-tank supply with a vegetable oil pre-heater to sufficiently 'thin' the oil.

Even if fresh vegetable oil is used in such a generator, it's arguable whether the energy is 'renewable' in that the carbon burned is absorbed again when the replacement batch of oil-bearing plants are grown. However, there are with the use and exploitation of inappropriate land for the growth of all types of bio-fuel, and the impact this has on Third World food prices. Literally, food for thought?

Our own renewable energy installation is equipped with a small vegetable oil generator from the German company Alternative Technology Group GmbH (ATG). This 2.8kW, air-cooled generator, named the 3SP, is the smallest in the ATG range, and is mounted on a steel frame to raise it to a height required by the installation. The frame is mounted on heavy rubber buffers (actually Land Rover bump-stops), with rubber pads between generator and stand.

There are many other vegetable oil compatible generators in the ATG range, right up to 1mW, with many other sizes in-between. These soundproofed generating sets are delivered ready-to-operate, and can be started at the push of a button or the turn of a key.

The supplementary exhaust was fabricated from a car exhaust, cut to length as required and supported above the ground. Note the stainless steel plate fitted permanently beneath the exhaust to prevent any vegetation from growing up around the exhaust, and creating a fire risk in summer.

The ATG generator, having been fixed to its steel frame, was then screwed to the wall to prevent the possibility of any movement from vibration. All that remained was for an electrician to make the connections to our renewable energy house supply, and it was good to go! (There are 230V and 12V take-offs from the generator itself, of course.)

All air-cooled generators are noisier than water-cooled ones, and short, stubby exhausts tend to be noisier than larger exhaust systems. I decided to modify the exhaust on my ATG generator by welding a flange to the exhaust outlet pipe.

Anaerobic digestion (AD)

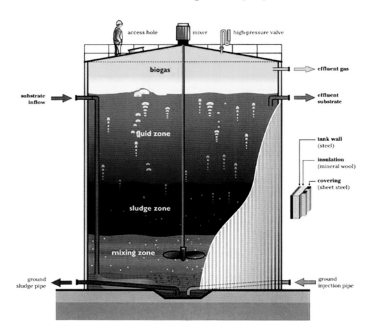

Where the exhaust was due to pass through a wooden wall in an outbuilding, the timber was cut away to within 200mm (8in) of the exhaust in all directions, and a metal plate made up in its place.

This is the City of London's first anaerobic digestion (AD) plant, from Eco Food Recycling Ltd. AD is a process by which plant and/or animal materials (biomass) are broken down by micro-organisms in the absence of air. Here's how it works –

1. Biomass* is placed inside a sealed tank (digester).
2. Mostly methane gas (biogas) is released as natural micro-organisms digest the biomass.
3. The gas can be used to generate heat and power, reducing greenhouse gas emissions.
4. The 'waste product' (digestate) is nutrient-rich; an excellent fertiliser.
*Such as crops and crop 'waste,' food waste, slurry and manure. It also keeps this waste from landfill. Biogas can even be refined to pure methane by removing other, unwanted gases, allowing it to be used in the mains gas grid or as road fuel.

Wave power

The UK has wave power levels that are among the highest in the world. East-facing sites in the UK are unsuitable but the south west has very good wave energy resources. A report by the Carbon Trust concludes that wave and tidal power could eventually provide around 20 per cent of the UK's electricity needs. (Courtesy Pelamis Wave Power Ltd)

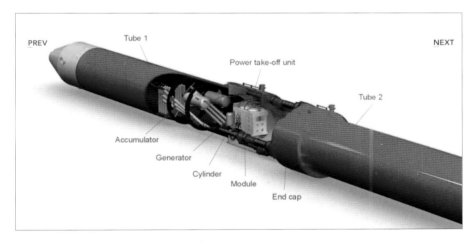

Of course, the sea is a very harsh place to site technology. The main problems are –

• The need to generate power over a wide range of wave sizes.
• Being able to withstand the largest and most severe storms.
• Other marine problems such as algae, barnacles and corrosion.

Wave energy has the potential to provide as much renewable energy as the wind, but in the short-term initial setup costs of marine energy are high, and extensive research and development are required.

Chapter 10
Tools, equipment and safety basics

Although this manual explains how the specialists carry out work for you, there will always be those who are sufficiently skilled and well qualified to do at least some of the work themselves, which could involve anything from basic home DIY through to fairly major construction projects associated with preparation for the fitting and assembly of renewable energy systems. This chapter is about how to choose the most suitable safety equipment and tools for working on your home. Product adverts won't tell you the whole story – it pays to dig a little deeper – and the following offers a few tips and suggestions …

SAFETY
BUYING BRAZING AND SOLDERING EQUIPMENT
CHOOSING A MIG WELDER
GETTING THE GAS
SELECTING A PLASMA CUTTER
CHOOSING A COMPRESSOR

In general, go for a top-notch manufacturer and avoid the high street 'specials' which are invariably of the cheap 'n' nasty variety, and try to match the compressor's capacity to your needs. But be aware that cheap compressors usually quote high (fictional?) CFM outputs. These are only approachable by running the motor and pump at excessive speeds. They won't last – and they'll be very noisy!

Different types of fire extinguisher are intended for different kinds of fire. You won't go wrong if you follow the online advice specifically for your country, such as that for the UK at www.firesafe.org.uk.

This is a car-sized battery: most lead acid batteries used for home energy storage are much larger and heavier, although the risks from acid spillage and fumes are exactly the same. Always wear rubber gloves, eye protection, and protective clothing when handling batteries, and follow manufacturer handling and installation instructions to the letter. There are more risks associated with storage batteries than many people realise!

Cheap safety gear is great ... but only if you don't care whether or not it works! Since the objective is to protect your eyes, your lungs, your hearing, and your health in general, this is one area where you really shouldn't cut corners. I invariably use Würth equipment and materials, recognised by many trades as among the best you can buy.

Similarly, you'd be a mug not to have the recommended number of both smoke and carbon monoxide alarms in your home, which will be even more at risk than usual while work inside it is going on.

Plumbing and major construction work will be easier with the benefit of a gas torch. This Bernzomatic self-igniting blow torch is actually being used for braze-welding aluminium …

CONSUMABLES AND HAND TOOLS
It's not commonly known that Würth

... with a Durafix Easyweld aluminium welding, brazing and soldering kit. It can cope just as well with regular copper pipe soldering.

You won't be able to carry out serious fabrication work without a MIG welder. Don't think that all are equal, however. An inverter-based welder will be lighter and (usually) produces smoother welds than transformer-based sets. My choice is the i-Tech MIG170 from Inverter Fusion. It's built to work for its living, so can be depended on for full restoration projects! The heavy-duty cycle means this machine won't keep cutting out if worked hard. There's enough max power for up to 5 or 6mm of weld penetration, while, unusually it also goes right down to 25A, so you can weld thinner metal without blow-through.

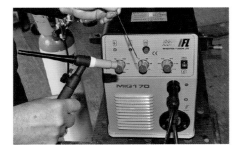

You can also add a TIG attachment for welding even thinner steel. You have to fit the torch to the positive terminal on the i-Tech MIG170 and flick the position switch. The gas must be pure argon because CO_2 destroys the tungsten tip.

The position switch is also flipped 'up' when the set is used as a conventional arc stick (or gasless MIG) welder. You can't usually MIG weld outside because wind disperses the shielding gas. But you can stick weld, or even use gasless wire with the i-Tech MIG170. So versatile!

Paying rental on a gas cylinder for occasional use, or using disposable welding cartridges adds up to a lot of cash! Alternatively, there's HobbyWeld's rent-free, refillable, welding gas cylinders. You buy a cylinder outright (about the cost of a decent tyre), and when you need a refill, just order a delivery or take the cylinder to your local stockist. Oxygen and acetylene also available.

When cutting through metal with plasma there's less distortion or collateral damage, reduced fire risk, lower running costs, less edge cleaning, and it's much quicker, as our test showed. But be aware that the cheapest plasma cutters won't cut as cleanly as this superb i-TECH cutter from Inverter Fusion. You'll need an air compressor producing 4.5 to 5.0 bar pressure, and a decent sized tank (see the ABAC compressor, here) – but no other gas. I selected an i-TECH CUT40 Inverter Plasma Cutter and here's what to look for –

1. 230V, single-phase current.
2. A decent duty cycle – this has 60%.
3. At maximum thickness, cuts tend to be messy. This 40 amp tool cuts steel up to 12mm thick, while a 25 amp tool might manage 6mm. Maximum quoted cutting thickness is invariably for mild steel; it'll be less for aluminium and stainless steel (neither of which can be cut with oxy-acetylene, by the way).
4. DON'T buy a cheap tool that requires you to 'strike' an arc. Buy one where you simply push a button to start the arc.
5. Two or three amperage positions are not enough. Best to have a variable control.
6. Look for a machine with integral regulator and pressure gauge.

When working out what you will need in a compressor, my advice is to go for mains voltage, single phase power; single-stage pump belt drive for quieter operation and greater longevity. Do you want a cheap, use-it-and-chuck-it compressor, or one that will last? My choice is an ABAC A29B 90 CM3. It's belt drive, 3hp, with a 90 litre tank, and produces 11.2 cubic feet per minute (CFM). This is a high-end or semi-pro machine (lower horsepower and smaller tank capacity versions also available from ABAC, as well as larger, of course!). TIP: In order to compare manufacturers, you often need to convert between differing mixtures of metric and imperial units.

also produces top quality hand tools under the Zebra label: an enormous range of German, top quality tools with a much longer life span than cheaper alternatives. As Würth says, 'Every product is subjected to endurance tests for days and sometimes weeks at our testing laboratories. These tests are always practice-oriented, and are usually much tougher than practical use itself.' In my experience with these tools, it shows.

This ABAC compressor can power many tools, by adjusting the pressure, including (low to high) –

Pneumatic (air) tool	Air consumption (CFM)
Spray gun (20-45psi)	0.5-3.5
Tyre inflator	2
Dust gun	3
Body polisher, drill or impact driver (½in)	4
Random orbit sander	6-9
Sand blaster	10-100s

PLENTY OF POWER

Many heavy-duty tools, such as welders and compressors, can test the limits of a UK 13A domestic power supply. The best solution is to have a dedicated 16A supply fitted. Here, you can see a stand-alone circuit breaker (left) for those without any spaces in their main supply board, plus a 16A socket and plug. UK law stipulates they must be connected by a qualified electrician.

SELECTING ELECTRIC CORDLESS TOOLS

For convenience and safety, I hardly ever use corded tools, but how to determine which are the best ones? In my case, I saw a cabinet-maker named Matt (seen working in this manual) using his Makita kit, and was so impressed I went out and got my own; five years on, most of my power tools are Makita. In front of my well-used Makita tools are the newer versions, with brushless motors giving 40% better battery life, and an invaluable battery meter on the drill (inset).

Here's Matt using a Makita 18V impact driver to fit a towel rail radiator in the bathroom, shown being installed in this manual. With a hammer action and maximum fastening torque of 160Nm, it releases all sorts of seized fixings. When fitted with a socket adapter, it swiftly zips bolts, screws and nuts on and off.

Appendices & index

Please note that all of the following are UK companies only because a listing of worldwide sources would fill a book! However, similar companies, organisations and information sources will be found in other countries.

ATAG Heating UK Ltd, Unit 3, Beaver Trade Park, Quarry Lane, Chichester, West Sussex PO19 8NY. Tel: 01243 815770 www.atagheating.co.uk

Baxi, Brooks House, Coventry Road, Warwick CV34 4LL. Tel: 0843 770 6599 www.baxi.co.uk

Bright Green Energy Ltd, Stable Units, Holmshaw Farm, Layhams Road, Keston BR2 6AR. Tel: 01959 570 728 www.brightgreenenergy.co.uk

Clearview Chimneys, trading As Michael Oliver Ltd, Rhos Cottage, Black Hill, Clun, Shropshire SY7 0JD. Tel: 01588 640910 www.clearviewchimneys.co.uk

Clearview Stoves, Dinham House, Ludlow, Shropshire SY8 1EU. Tel: 01584 878100 www.clearviewstoves.com

Co-op Energy. Freephone: 0800 954 0693 www.cooperativeenergy.coop

DARE (Devon Association for Renewable Energy), 12a The Square, North Tawton, Devon EX20 2EP. Tel/fax: 01837 89200 www.dare.btck.co.uk

Department of Trade and Industry www.dti.gov.uk

Durafix, for information on where to purchase Durafix, see www.durafix.com

Eco Food Recycling Ltd, 36 Wayside Road, St Leonards, Ringwood, Hants BH24 2SJ. Tel: 01202 873967 www.ecofoodrecycling.co.uk

Eco2Solar Ltd, Unit 8, John Samuel Building, Arthur Drive, Hoo Farm Industrial Estate, Kidderminster, Worcs DY11 7RA. Tel: 01562 977977 www.eco2solar.co.uk

Eco-Eye Ltd, The Modern Moulds Business Centre, 2-3 Commerce Way, Lancing, West Sussex BN15 8TA. Tel: 01903 851910 www.eco-eye.com

Economical Energy Solutions Ltd, The Old Pump House, New Street, Upton-upon-Severn WR8 0HP. Tel: 01684 594981

www.economicalenergysolutions.co.uk

Efficient Energy Centre, 2 Harrow Road, Plough Lane, Hereford HR4 0EH. Tel: 01432 356633 www.efficientenergycentre.co.uk

Energy Action (Scotland) is a national charity aiming to end fuel poverty and ensure warm, dry homes for all. Tel: 0141 2263064 www.eas.org.uk

Energy Saving Trust, 21 Dartmouth Street, London SW1H 9BP. Tel: 0300 123 1234 www.energysavingtrust.org.uk

Energy Share, www.energyshare.com

Euroheat (HBS) Ltd, Unit 2, Court Farm Business Park, Bishops Frome, Worcester WR6 5AY. Tel: 01885 491100 www.euroheat.co.uk

Genius Roof Solutions, Unit F23, Coppull Mill Enterprise Centre, Mill Lane, Coppull, Lancs PR7 5BW. Tel: 01257 793113 www.geniusroofsolutions.com

Geologic Boreholes, Unit 2, Moor View Industrial Estate, Whimple, Exeter, Devon EX5 2QT. Tel: 01404 822032 www.geologicboreholes.co.uk

Grants (UK) – for available UK government grants for home heating and insulation schemes, see www.gov.uk/warm-front-scheme

Jeremy Hayward (plumber), Clifton-upon-Teme, Worcestershire. Tel: 01886 812494 Mobile: 07712 254451

Home Heat Helpline – advice for people in the UK who are worried about paying their fuel bills and keeping warm, with special help for low-income households. Freephone: 0800 336 699 www.homeheathelpline.org.uk

Ice Energy Heat Pumps Ltd, Unit 2, Oakfield House, Eynsham, Oxon OX29 4TH. Tel: 0808 145 2340 www.iceenergy.co.uk

JW Solar Solutions, Justin Walters. Tel: 01905 821691 www.jwsolarsolutions.co.uk

Klober Ltd, Unit 6F, East Midlands Distribution Centre, Short Lane, Castle Donington, Derbyshire DE74 2HA. Tel: 01332 813 050 www.klober.co.uk

Charles Landau, www.solarpaneltilt.com

Leading Edge Turbines Ltd, Skyrrid Farm, Pontrilas, Hereford HR2 0BW. Tel: 0845 652 0396 www.leturbines.com

Makita UK Ltd, Michigan Drive, Tongwell, Milton Keynes, Bucks MK15 8JD. Tel: 01908 211678 www.makitauk.com

Marflow Hydronics (Filterball®) – see Speedfit

Microgeneration Certification Scheme (MCS), Tel: 0207 090 1082 www.microgenerationcertification.org

Mitsubishi Electric Europe BV, Travellers Lane, Hatfield Herts AL10 8XB. Tel: 01707 282880 www.livingenvironmentalsystems. mitsubishielectric.co.uk

National Energy Action (England and Wales) – a charity that helps people on low incomes to heat and insulate their homes. Tel: 0191 261 5677 www.nea.org.uk

Newark Copper Cylinders, Brunel Drive, Northern Road Ind. Estate, Newark, Nottinghamshire NG24 2EG. Tel: 01636 678437 www.newarkcoppercylinder.co.uk

North American Renewable Energy Directory, www.nared.org

Pedmore Plumbing Heating & Gas, Peter Barnett, 5 Wychbury Road, Pedmore, Stourbridge, West Midlands DY9 9HP. Tel: 01384 356033 Email: pedmore@ mac.com

Pelamis Wave Power Ltd, 31 Bath Rd, Leith, Edinburgh EH6 7AH Scotland. Tel: 0131 554 8444 www.pelamiswave.com

Plasson UK Ltd, Plasson House, 27 Albert Drive, Burgess Hill, West Sussex RH15 9TN. Tel: 01444 244446 www.plasson.co.uk

The Renewable Energy Home Handbook

Power Predictor 2, from Eco Ark Ltd, Tressa, Legion Lane, Par, Cornwall PL24 2QR. www.ecoark.co.uk

Renusol GmbH, Piccolominstr. 2, 51063, Cologne, Germany. Tel: +49 221 788 7070 www.renusol.com

RH Nuttall Ltd, Great Brook Street, Birmingham B7 4EN. Tel: 0121 359 2484 www.rhnuttall.co.uk

Screwfix Direct, Freepost, Yeovil, Somerset BA22 8BF. Freephone: 0500 41 41 41 www.screwfix.com

SimplyLED, Unit 4, Leeds West, Gelderd Lane, Leeds LS12 6AL. Tel: 0845 459 8010 www.simplyled.co.uk

SMA Solar UK Ltd, Unit 2B, Gemini Building, Sunrise Parkway, Linford Wood, Milton Keynes MK14 6NP. Tel: 01908 304850 www.sma-uk.com

Solfex Energy Systems, Energy Arena, 3-5 Charnley Fold, Bamber Bridge, Preston, Lancashire PR5 6PS. Tel: 01772 312847 www.solfex.co.uk

Speedfit, Britannia House, Austin Way, Hamstead Industrial Estate, Birmingham B42 1DU. Tel: 0121 358 2012 www.marflowhydronics.co.uk

StormDry, Safeguard Europe Ltd, Redkiln Close, Redkiln Way, Horsham, West Sussex RH13 5QL. Tel: 01403 210204 www.safeguardeurope.com

The Environment Agency, National Customer Centre, PO Box 544, Rotherham S60 1BY. Tel: 03708 506 506 www.gov.uk

Trojan Battery Co, www.trojanbatteryre.com

Ubbink (UK) Ltd, Unit 33, Liliput Road, Brackmills Ind Est, Northants NN4 7DT. Tel: 01604 433000 www.ubbink.co.uk

Würth UK Ltd, 1 Centurion Way, Erith, Kent, DA18 4AF. Tel: 03300 555 444 www.wurth.co.uk

GLOSSARY

Absorber plate   A dark surface that absorbs solar radiation and converts it to heat; a component of a flat-plate solar collector.

AC   Abbreviation for alternating current: a current that periodically reverses direction – measured in cycles per second (hertz, Hz).

Accumulator   A cell, solar pond, thermal mass, or other device that stores energy, such as a battery.

Active power (also known as 'real power' or simply 'power')   Active power is the rate of producing, transferring or using electrical energy. Measured in watts and often expressed in kW or mW.

Agreed capacity   An agreed amount of electrical load for a property, as stated in the property's Connection Agreement with the local distribution network operator (DNO).

Air source heat pump (ASHP)   The heat pump absorbs heat from outside air and, in heating mode, transfers this to where heat is required. In cooling mode the heat pump absorbs heat from the space to be cooled, and vents this to the outside.

Alternator   A generator which changes mechanical energy into AC electrical energy.

Amp   Abbreviation for ampere: a unit that measures the rate of flow of an electrical current.

Anemometer   A device that rotates with the wind to measure wind speed.

Automatic meter read (AMR)   The term given to a system that provides automatic meter readings remotely, transferring data to a billing system.

Automatic tracking   A device that allows solar collectors to 'track' or follow the sun during the day.

Auxiliary generator   A small, enginedriven generator that supplements a renewable energy power source.

Balancing mechanism   Used by the National Grid to balance electricity supply and demand.

Biodiesel   The biofuel substitute for diesel. Usually processed/refined from oilseed-based crops. NOT the same as SVO (straight vegetable oil).

Bioethanol   The biofuel substitute for petrol (gasoline) – usually derived from cereal-based crops.

Biogas   The biofuel substitute for natural gas. Usually derived from organic waste materials, including animal waste and other waste via anaerobic digestion.

Biomass   Solid fuel derived from plant and organic matter which is used to create heat or generate electricity. While not entirely pollution-free, this contributes little to global warming.

Brine   A heavy salt solution, mistakenly described as the fluid with anti-freeze that used in a heat pump or refrigeration system.

Building energy rating (BER)   UK legislation that rates energy efficiency of new and existing, non-residential buildings.

Calorific value (CV)   Amount of heat generated by a specified amount of gas, measured in joules per kilogram.

Carbon capture and storage (CCS)   The process of capturing carbon that is emitted from energy production and storing it, to reduce the amount of CO_2 emitted into the atmosphere.

Carbon dioxide (CO_2)   An inert, non-toxic gas produced from decaying materials, plant and animal life respiration, and combustion of organic matter, including fossil fuels.

Carbon dioxide equivalents (CO_2e)   The internationally recognised way of expressing the amount of global warming of a particular greenhouse gas in terms of the amount of CO_2 required to achieve the same warming effect over 100 years.

Carbon footprint   The total emission of greenhouse gases (in carbon equivalents) from whichever source is being measured.

Carbon labelling   Consumer measurement of the amount of embedded carbon in a product.

Carbon neutral   Offsetting greenhouse gas emissions produced by carrying out an equivalent amount of carbon saving in another area to make the producer 'carbon neutral.'

Carbon offsetting   Offsetting greenhouse gas emissions by purchasing credits from others through emissions reductions projects, or carbon trading schemes.

Carbon sink   An absorber of carbon dioxide – oceans and forests are natural carbon sinks.

Carbon Trust   An independent, non-profit-making company set up by the UK government with support from businesses to encourage and promote the development of low carbon technologies.

Celsius (sometimes called Centigrade)   The international temperature scale in which water freezes at 0 [degrees] and boils at 100 [degrees].

CHP   A combined heat and power machine whereby waste heat from the process of generating electricity is used to create energy (usually electrical or thermal).

Circuit   A joined-up series of electrical conductors, wires and components that allow an electrical current to flow through it.

Circuit breaker   A device (switch; often automatic) that protects a circuit from power surges by preventing power flow.

Collector plate   A device to trap solar radiation and convert it to usable heat.
Combined heat and power (CHP)   A system whereby fuel used to produce electrical (or mechanical) power simultaneously recovers useful thermal energy for heating.

Compound parabolic collector   A type of solar collector using parabolic reflectors.

Conduction   Heat transfer between a hotbody and a cold body.

Conductor   A substance that allows an electrical current to pass through it easily.

Connection agreement   A document which states the agreed capacity for a property with the local (UK) distribution network operator (DNO).

Convection   The natural convection of heat through fluid or gas. Warm, less dense fluid rises, whilst cold, dense fluid sinks.

Coefficient of performance (CoP)   Energy output to electricity consumption ratio, used mainly as a way of measuring heat pump efficiency. Similar to seasonal performance factor (SPF) and seasonal coefficient of performance (SCOP).

Deemed amount   A contract with an electrical supplier with a defaulted rate for supply until a customer requests a fixed price for a fixed period. OR a payment made for electricity supplied to the grid by a microgenerator, calculated on the potential output of the equipment installed but not measured.

DC   Abbreviation for direct current: an electrical current that flows in one direction only in a circuit. Batteries and fuel cells produce direct current.

Distributed generation   Eectricity generation, usually on a relatively small scale, that is connected to the distribution networks rather than directly to national transmission systems.

Distribution losses (Dloss)   The loss of distributing power through grid network wires.
Distribution network operators (DNO)   Companies which are responsible for operating the grid networks, connected to electricity consumers.
Distribution system   The local poles, wires, transformers, substations, and other equipment used to deliver electricity to end-use consumers. Also 'grid.'

Double-pole switch   Switches (disconnects) both live and neutral wires.

Efficiency   The ratio (usually a percentage) of energy output to energy input.

Electric current   The rate at which electricity flows through an electrical conductor, usually measured in amperes (amps).

Electrical cell   A device (usually within a battery) which produces or stores electricity.
Electricity metre   A device that measures the amount of electricity used.

Energy efficiency   Ratings for new appliances or building improvements that reduce energy used, whilst maintaining the same performance benefits.

Energy performance of building directive (EPBD)   The principle objective of this EU directive is to promote improvement of energy performance of buildings within the EU via cost-effective measures.

Energy saving trust (EST)   The UK's EST is an independent non-profit organisation, set up and largely funded by the government to manage a number of programmes to improve energy efficiency, particularly in the domestic sector.

Energy source   The primary source of energy, whether from fossil fuels or renewables.

Environment agency   The leading public body for protecting and improving the environment in England and Wales.

Evacuated tube collector   A solar device that uses a vacuum inside a glass tube to

insulate the absorber plate.

Export metering   A device for measuring the amount of exported electricity.

Fahrenheit   The temperature scale in which water freezes at 32 [degrees] F and boils at 212 [degrees] F.

Flat-plate solar collector   A device which uses an absorber plate to collect solar radiation.

Flex   Electrical wire covered with insulating material.

Forest Stewardship Council (FSC)   An independent, non-governmental, not-for-profit organisation set up to respond to concerns about global deforestation.

Fossil fuel   An energy source formed in the Earth's crust from decayed organic material. Common fossil fuels are oil, coal, and natural gas.

Fuel cell   These produce electricity from hydrogen and air, with water as the only emission.

Fuse   A thin wire inside a protective case that melts and breaks if the flow of electricity becomes too powerful, thus protecting electrical appliances.

Generation   Electricity production.

Generator   A machine that converts mechanical energy into electricity.

Geothermal energy   Energy generated from heat stored beneath the earth's surface. (NB NOT the same as ground source energy.)

Giga watt (gW)   1 gW = 1000 MW (mega watt).

Global warming/effect   The greenhouse gas effect caused – the majority of scientists agree – by human activities such as burning fossil fuels and other industrial processes, which release greenhouse gas into the atmosphere.

Greenhouse gas/effect   The way gases in the earth's atmosphere trap heat. A build-up of these gases, especially carbon dioxide, is thought to cause global warming.

Ground source heat pump (GSHP)   A heat pump that uses the earth's natural heat storage ability and/or groundwater to heat and/or cool a building.

Head   The vertical distance from the point where water enters an intake to the point where water leaves a device. Generally measured in meters or feet. Head x flow is a measurement of potential power.

Heat exchanger A device, such as a coiled copper tube immersed in a tank of water, which is used to transfer heat from one fluid to another via a separating wall.

Heat pump A electro-mechanical device that transfers heat from a heat source to a destination in order to extract more energy than is used in running the heat pump.

Heat rate Energy input per unit of time, usually expressed in kW/h or BTU/h.

Heat recovery Collection and re-use of heat from a building that would otherwise be lost.

HV High Voltage (11,000 Volts or above).

Hybrid power system A system, such as one might find within a home, that derives its heating, lighting, and other energy from several interconnected sources, one or more of which would usually come from renewable energy.

Hydroelectricity Using the force of falling water to generate electricity by turning turbine blades.

Import Where a site consumes electricity as opposed to generating and exporting power.

Induction motor A common type of motor which, driven by a windmill or turbine, provides alternating current (AC) electricity.

Insolation The process by which energy from the sun reaches the earth's surface. Insolation is generally measured in BTU/square feet (meters)/day, and we frequently talk about the 'amount' of insolation utilised.

Insulator A material that reduces or halts the flow of electricity.

Inverter A device that converts direct current (DC) to alternating current (AC) (or vice versa).

Kilowatt (kW) A measure of instantaneous power. One kilowatt equals 1000 watts.

Kilowatt/hour (kW/h or kWh) One kilowatt of electricity usage per hour.

kWe, kWm, kWt kilowatts of electricity, mechanical power and thermal energy, respectively All the same quantity of power but used to quantify the 'type' of energy being produced by a generator.

Load The power carried by a particular circuit.

Load factor Measures the relationship between unit consumption and maximum demand, and is the percentage capacity utilisation figure of a site's power consumption. Expressed as the total number of units of consumption, divided by maximum demand, divided by number of hours, multiplied by 100.

Mains electricity Power supplied to homes from the grid.

Marine generation (tidal and wave) Similar to wind turbines, except that the process uses underwater current, and/or tides.

Mega watt (MW) A measure of power: one million watts.

Meter asset provider (MAP) The party responsible for ongoing provision of the meter installation at that metre point. The MAM might be the owner of the meter, or could lease or rent the meter from a third party.

Meter operator (MOp) The organisation appointed to maintain metering equipment.

Meter serial number The unique number stamped on the front of the meter.

Micro-CHP CHP (as above), but in very small scale, typically below 5kW electrical output, (eg in residential and commercial sectors). It is likely to operate in place of a domestic central heating boiler.

Micro-generation Small-scale generation of energy, usually from renewable sources, for example, solar panels or domestic wind turbines.
MPAN See Supply Number (S-number).

MWh mega watt hour, one thousand kWh A 1 MW power-generating unit running for 1 hour produces 1 MWh of electrical energy.

National Grid The UK's National Grid owns the main transmission systems, and is responsible for transmitting electricity from the generator and through the National Grid, before it is fed into distribution networks.

Passive solar heating Solar heating a building by use of architectural design without the aid of extraneous equipment.

Peak demand Point of maximum electricity demand on the national system.

Peak watt Unit used for the performance

rating of photovoltaic converters. A one peak watt system will deliver one watt at the specified working voltage under peak solar irradiation.

Photovoltaic array A number of photovoltaic modules, connected in series and/or parallel so as to provide desired power and voltage.

Photovoltaic cell Solar energy device that changes light into electrical energy.

Photovoltaics (PV) The direct conversion of solar radiation to electricity by the interaction of light with the electrons in a semiconductor device or cell.

Power line Electrical wires that carry electricity from the point of generation to the point of use.

Pylon A large, metal tower that carries very high voltage power lines.

RCD (residual current device) A life-saving device designed to prevent fatal electric shock if something live – such as a bare wire – is touched. Can also provide some protection against electrical fires. RCDs offer a much greater level of personal protection than ordinary fuses and circuit-breakers.

Renewable energy A term used to describe energy produced using naturally-replenishing resources, which includes solar power, wind, wave and tide, and hydroelectricity. Wood, straw and waste are often called solid renewables; landfill and sewerage gas are known as gaseous renewables.
Renewable obligation

(RO) The main UK government market mechanism to support renewable energy. It is an obligation for all electricity suppliers to source a certain part of electricity sales from accredited renewable sources.

Renewable Power Association (RPA) A trade association open to all companies supportive of the UK renewable energy industry.

Renewables Energy sources that are either inexhaustible (solar, wind) or replenished over a short period of time. In the USA: renewable energy credits (RECs), green tags or tradable renewable certificates (TRCs).

Renewables obligation certificate (ROC) Eligible renewable generators receive ROCs for each MWh of electricity generated.

Smart metering The ability to remotely

read electricity meters. Data is more reliable, and more accurate bills are produced as a result. Information is often available 'real-time' to the customer.

Solar absorption   Absorption of solar radiation by a material.

Solar altitude   The sun's angle above the horizon, as measured in a vertical plane.

Solar Azimuth   The horizontal angle between the sun and due south in the northern hemisphere, or between the sun and due north in the southern hemisphere.

Solar collector   A device that gathers and accumulates solar radiation to produce heat via a solar absorber plate.

Solar energy   Electromagnetic radiation generated by the sun which may be converted to useful forms of energy.

Solar fraction   The amount of energy provided by the solar technology divided by the total energy required

Solarirradiance   The total amount of solar radiation striking a given area.

Solar panels   Solar thermal (solar water heating) panels, used to heat water AND solar electric – photovoltaic (PV) systems – used to convert light to electricity.

Solar pump   Operates on solar energy,

either by a photovoltaic process or by a thermal system in which a fluid heated by the sun drives a turbine or piston that powers the pump.

SCOP and SPF Re   Heat pumps – see CoP.

Stirling engine   An external combustion engine in which air is alternately heated and cooled using relatively low levels of energy to drive a piston up and down.

Sub station   A part of the National Grid that contains transformers which increase or decrease the voltage of an electric current.

Supplier   An organisation authorised by a supply licence to supply electricity or gas to the National Grid network.

Supply number or S-number (also known as MPAN – meter point administration number)   A unique number identifying the distribution company and location of the metering point.

SVO (straight vegetable oil)   Vegetable oil used, unprocessed, as fuel.

Therms   A unit of energy measurement: kWh x 29.3071.

Tiltangle   Angle at which a solar collector is tilted upward from the horizon for maximum light or heat collection.

Tracked array   A photovoltaic array which follows the path of the sun across the sky.

Transformer   Equipment used to increase or decrease the voltage of an electric current.

Transmission (lines)   Transfer of electricity at high voltage from the power stations across the UK through interconnected electric lines, known as the grid system

Turbine   A device that converts the energy in a stream of air or fluid to mechanical energy.

U-value   A measurement of the amount of heat flowing in or out of a building or room in one hour, under constant conditions, where there is a one degree difference in temperature between the air inside and outside.

Voltage   A unit used to measure the electromotive force of an electric current.

Watt   The unit rate at which work is done in an electrical circuit. One watt equals one joule of work per second.

Wind power   The conversion of wind energy into more useful forms, usually electricity, using wind turbines: generators with blades.

More titles from Veloce Publishing

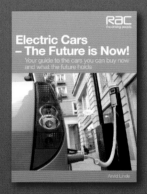

Electric Cars

What if we all had to say goodbye to petrol cars tomorrow? Would you be ready? This book will help you find out. With a concise catalogue covering the best production models and the most promising prototypes, this book is the definitive guide to the future of motoring.

ISBN: 978-1-845843-10-6
Paperback • 21x14.8cm • £12.99* UK/$24.95* USA • 128 pages

Dorset from the sea

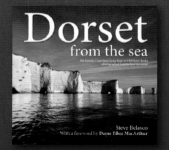

The photographs in this book have been taken entirely from the sea by sailor and marine photographer Steve Belasco, who has cruised the area in small boats for over 20 years.

His love of Dorset's waters, those who enjoy them, and the people and creatures that depend on them, comes shining through.

ISBN: 978-1-845847-62-3
Hardback • 25x25cm • £24.99* UK/$45* USA

France: the essential guide for car enthusiasts

Whether you prefer vintage models, or the latest sports cars, you'll find 200 ideas for places to see and events you can take part in inside this unique guide to France. Covering everything automotive, from museums and concours d'élégance, to motorsport events and track days, this book is packed with useful information and essential data.

ISBN: 978-1-845847-42-5
Paperback • 21x14.8cm • £14.99* UK/$24.95* USA • 248 pages

For more info on Veloce titles, visit our website at www.veloce.co.uk • email: info@veloce.co.uk • Tel: +44(0)1305 260068
* prices subject to change, p&p extra

eBooks from Veloce Publishing

From technical manuals, to photo books, to autobiographies, Veloce's range of eBooks gives you the same high-quality content, but in a digital format tailored to your favourite e-reader. Whether it's a technical manual, an autobiography, or a factual account of motorsport history, Veloce's ever expanding eBook range offers something for everyone.

See our website for information on our full range of eBook titles.

www.veloce.co.uk

Available on the iBookstore

kobo available for kobo ereaders

Available for amazonkindle

Available for nook by Barnes & Noble

W ebook available from Waterstones

ebook available from Google play

INDEX

188